EVOLUTIONARY THEORY IN SOCIAL SCIENCE

THEORY AND DECISION LIBRARY

General Editors: W. Leinfellner and G. Eberlein

Series A: Philosophy and Methodology of the Social Sciences
Editors: W. Leinfellner (Technical University of Vienna)
G. Eberlein (Technical University of Munich)

Series B: Mathematical and Statistical Methods
Editor: H. Skala (University of Paderborn)

Series C: Game Theory, Mathematical Programming and Mathematical
Economics
Editor: S. Tijs (University of Nijmegen)

Series D: System Theory, Knowledge Engineering and Problem Solving
Editor: W. Janko (University of Vienna)

SERIES A: PHILOSOPHY AND METHODOLOGY OF THE SOCIAL SCIENCES

Editors: W. Leinfellner (Technical University of Vienna)
G. Eberlein (Technical University of Munich)

Scope

This series deals with the foundations, the general methodology and the criteria, goals and
purpose of the social sciences. The emphasis in the new Series A will be on well-argued,
thoroughly analytical rather than advanced mathematical treatments. In this context,
particular attention will be paid to game and decision theory and general philosophical topics
from mathematics, psychology and economics, such as game theory, voting and welfare
theory, with applications to political science, sociology, law and ethics.

EVOLUTIONARY THEORY
IN
SOCIAL SCIENCE

Edited by

MICHAEL SCHMID

University of Augsburg,
Department of Sociology, Augsburg, F.R.G.

and

FRANZ M. WUKETITS

University of Vienna,
Department of Philosophy of Science, Vienna, Austria

D. REIDEL PUBLISHING COMPANY

A MEMBER OF THE KLUWER ACADEMIC PUBLISHERS GROUP

DORDRECHT / BOSTON / LANCASTER / TOKYO

Library of Congress Cataloging in Publication Data

Evolutionary theory in social science / edited by Michael Schmid and Franz M.
Wuketits.
    p. cm.—(Theory and decision library. Series A, Philosophy and methodology of the
social sciences)
    Bibliography: p.
    Includes indexes.
    Contents: Basic structures in human action / Peter Meyer—Evolutionary models
and social theory / Michael Ruse—Evolution, causality, and freedom / Franz M.
Wuketits—Collective action and the selection of rules / Michael Schmid—Learning and
the evolution of social systems / Klaus Eder—Evolution and political control / Peter A.
Corning—Media and markets / Bernd Giesen—The self as a parasite / Richard Pieper.
    ISBN 90–277–2612–4
    1. Social evolution. I. Schmid, Michael, 1943– . II. Wuketits, Franz M.
III. Series.
GN360.E92 1987
573.2—dc 19   87–23584
                                                            CIP

Published by D. Reidel Publishing Company,
P.O. Box 17, 3300 AA Dordrecht, Holland.

Sold and distributed in the U.S.A. and Canada
by Kluwer Academic Publishers,
101 Philip Drive, Assinippi Park, Norwell, MA 02061, U.S.A.

In all other countries, sold and distributed
by Kluwer Academic Publishers Group,
P.O. Box 322, 3300 AH Dordrecht, Holland.

TABLE OF CONTENTS

PREFACE   ix

PETER MEYER / Basic Structures in Human Action. On
   the Relevance of Bio-Social Categories for Social
   Theory   1

      I.   The Problem   1
     II.   Some Preconditions of Behavioural Patterns   2
    III.   Taking Phenotypes Seriously: Critical Remarks
           on Sociobiology   5
     IV.   Secondary Type Explanations do not Explain away
           Primary Type Explanations   9
      V.   Biosociology: A Levels Model of Man   12
     VI.   The Incest Taboo: A Biosociological View   15
    VII.   The Human Biogram and the Role of Cultural
           Institutions   17
   VIII.   Conclusion   20
   Notes   21

MICHAEL RUSE / Evolutionary Models and Social Theory.
   Prospects and Problems   23

      I.   Introduction   23
     II.   Social Darwinism   24
    III.   Animal Sociobiology   28
     IV.   Human Sociobiology   31
      V.   The Evolution of Morality   33
     VI.   The Status of Morality   37
    VII.   Relativism ?   40

VIII. Relatives, Friends, and Strangers   42
  IX. Prospects   44
   X. Conclusion   47

FRANZ M. WUKETITS / Evolution, Causality and Human Freedom.
The Open Society from a Biological Point of View   49

    I. Introduction   49
   II. The Systems-Theoretic Approach to Evolution:
        Darwin and Beyond   51
  III. The Evolution of Man: Beyond Determination and
        Destiny   59
   IV. The Evolution of Man: Beyond Physicalism and
        Mentalism   62
    V. Evolution and the Open Society   67
   VI. Conclusion   75
  Notes   75

MICHAEL SCHMID / Collective Action and the Selection
of Rules. Some Notes on the Evolutionary Paradigm in
Social Theory   79

    I. On the Genesis of the Social Theory of
        Evolution   79
   II. The Logical Structure of a Theory of Structural
        Selection   83
  III. An Action-Theoretical Interpretation of the
        Theory of Structural Selection   85
   IV. The Heuristics of the Theory of Structural
        Selection   91
    V. Conclusion   98
  Notes   99

KLAUS EDER / Learning and the Evolution of Social
Systems. An Epigenetic Perspective   101

    I. Evolution and the Role of the Epigenetic System   101
   II. Epigenesis and Evolution in Sociological
        Theorizing   103
  III. Epigenetic Developments and Social Evolution   111
   IV. An Epigenetic Theory of the Formation of the
        State   115
    V. Conclusion   123
  Notes   125

PETER A. CORNING / Evolution and Political Control.
A Synopsis of a General Theory of Politics   127

    I. Introduction   127
   II. The Theoretical Problem   128
  III. Evolutionary Causation   130
   IV. Functional Synergism   132
    V. The Cybernetic Model   138
   VI. A General Theory of Politics   142
  VII. Some Theoretical Implications   157
 VIII. Conclusion   169

BERNHARD GIESEN / Media and Markets   171

    I. Introduction   171
   II. The Selectionist Program   172
  III. Money and Language: Two Models for General
        Media of Interaction   176
   IV. The Institutionalization of the Media Codes:
        Structural Requirements   177
    V. Communities, Hierarchies and Markets   179
   VI. Political, Socially Intergrative and Scientific

          Markets   183
VII.   Concluding Remarks: Media Between Inflation and
          Deflation   192
Notes   194

RICHARD PIEPER / The Self as a Parasite.
A Sociological Criticism of Popper's Theory of
Evolution   195

   I.   Introduction   195
  II.   Dualism, Trialism or Pluralism ?   198
 III.   Descarters' Problem   203
  IV.   Propensities as Collective Social Forces:
          Durkheim   209
   V.   The Self as a Parasite   214
  VI.   Epistemology and the Knowing Subject   220
Notes   222

BIBLIOGRAPHY   225

INDEX OF NAMES   255

INDEX OF SUBJECTS   261

PREFACE

In retrospect the 19th century undoubtedly seems to be the century of
evolutionism. The 'discovery of time' and therewith the experience of
variability was made by many sciences: not only historians worked on the
elaboration and interpretation of this discovery, but also physicists,
geographers, biologists and economists, demographers, archaelogists, and
even philosophers. The successful empirical foundation of evolutive
processes by Darwin and his disciples suggested Herbert Spencer's
vigorously pursued efforts in searching for an extensive catalogue of
prime and deduced evolutionary principles that would allow to integrate
the most different disciplines of natural and social sciences as well as
the efforts of philosophers of ethics and epistemologists. Soon it became
evident, however, that the claim for integration anticipated by far the
actual results of these different disciplines. Darwin's theory suffered
from the fact that in the beginning a hereditary factor which could have
supported his theory could not be detected, while the gainings of ground
in the social sciences got lost in consequence of the completely
ahistorical or biologistic speculations of some representatives of the
evolutionary research programm and common socialdarwinistic
misinterpretations. In the social sciences the influence of evolutionary
ideas was extensively narrowed by functional analysis which engaged in the
development of equilibrium theories whose logic allowed to describe
processes of reproduction but not of transformation, and consequently the
historical sciences succeeded in insisting upon the necessity of applying
completely independent methods of research leading away from other social
sciences. In absence of empirically realizable models the attempts to
develop an evolutionary philosophy came to nothing and merely were
suspended considering the breakthrough of logical positivism towards
evolutionary ethics had to face the still current argument that (normative
and ethical) questions of evaluation, in whose reply the academic
philosophy declared itself competent, could not be solved with reference
to the history of evolution.

Despite of these important objections the evolutionary research program succeeded in disentangling from these restrictions. Darwin's theory found a micro-biological basis in modern genetics, what led to a 'new synthethis' of the biological theories in the middle of the fourties and actually seems no more than marginally endangered. In the late fifties the evolutionary interpretation of social development emerged again, following the preliminary studies of social anthropology and social archaelogy that never had abandoned the contact to evolutionism. Later on micro-economics and the organizations theory remembered the importance and fruitfulness of selection theoretical models and quite recently macro-economics has taken up the at times hidden evolutionary heritage. And evolutionary epistemology as well as new approaches to an evolutionary understanding of ethics won new adherents.

This at first unexpected revitalization of evolutionary thinking was due to a number of extensive theoretical displacements. On the one hand it was pointed out unmistakably that the acceptance of evolutionary models and therewith of their basic idea of conceiving evolution as **selective process,** was not to be equaled logically with models of an individualistic competition caused by limited resources. On the contrary it could be shown that in the context of evolutionary theory the stabilization of forms of cooperation based on division of labor could also be explained and that, moreover, there were various forms of reciprocal association whose conditions for reproduction and transformation could be identified successfully. The examiniation of interrelations between these forms of association opened a rich heuristics that could be elaborated with the help of ecologic theories. Any ideological commitment can be avoided in the range of these theoretical attempts.

On the other hand an important lack of the classical paradigma of evolution could be eliminated. The antiquate model had always been confronted with the question, whether it was sufficient to explain the forming of structures and the increase of structural complexity only as the effect of external selections which allowed contingent variations to gain an advantage in reproduction. In this context neither the assumed existence of successful processes of replication nor the demand for the empirical proof of all sorts of systems able for evolution were contested. But there still was the problem, how the development and the increase in complexity of structures could be explained under the hypothesis of classical thermodynamics, whose entropy theorem logically excluded the forming of structures. As a matter of fact, nowadays this problem can be considered as solved if one remembers for one thing that modern **nonequilibrium thermodynamics** offers a model for the emerging of structures from the flow of energy, far from thermodynamic (and i.e. entropic) balance, and furthermore that the discovery of **recursive systems processes** offered a tenable explanation for the fact that due to the ability of a process of reproduction to react on its own results and distributions, what means being operatively closed, structures invariably emerge. Therewith these processes of structuration are no longer mysterical. These ideas can easily be associated with the classical theory on the importance of external selection. So the objection of scientists who wanted to eliminate the model of selection from scientific discussion on the occasion of its empirical deficiences, looses plausibility.

At the same time it became evident that models of linear causality were insufficient for a close analysis of self-organizing processes, but nevertheless it no longer has been possible to eliminate this lack by postulating doubtful theoretical entities. This may be omitted by realizing that feedback-causalities, recursive self-reference and hypercyclical forms of organization are sufficient to explain the inevitable appearance of at the same time emerging and self-reproductive levels of structures.

This theoretical development again offers the possibility to determine an integral scientific program enabling to combine the different aspects of structural self-organization, reciprocal exchange of energy and external selection and therefore allows to reconstruct evolutionary theory as an **universal theory of dynamic and self-transformative systems**. It provides a 'hard core' whose universality is shown by the possibility of being adopted also in social sciences. The following papers share the certainty that this demand for universality is for right. However, there are the most different ways of making evolutions theoretical suggestions for systematization profitable to the further development of Social Theory.

First of all, the remark that human acting is a natural phenomenon emerging from evolution has to be taken most seriously and it should be examined in which way its biogenetic presuppositions have an influence on the structure of social relations and cultural institutions. This is done by **Peter Meyer** in his paper 'Basic Structures in Human Action. On the Relevance of Bio-Social Categories for Social Theory'. In his argumentation he tries to show that structured emotions are the basis of every form of human acting, and how these patterns cause and limit the forms of possible institutionalization of social acting. It is demonstrated how on the basis of bio-grammatical suppositions of human action emerges a level of a socio-culturally mediated sociality on which selective influences opperate in the same way as on lower levels of hehaviourial structurization.

These arguments are completed in a most engaged way in **Michael Ruse's** essay on 'Evolutionary Models and Social Theory. Prospects and Problems' by pointing out that the moral convictions of man have biogenetic roots and that many conflicts in moral acting result from forgetting this fact and therewith failing to notice that evolutively developed morals gained their adaptive character in social and ecologic circumstances which nowadays no longer are evident. An exit from these limitations could only be found by realizing this fact and by being conscious of having to act in the context of a common moral heritage of the genus man, but in no way by postulating unfullfillable moral claims neglecting the mentioned roots of our moral conceptions.

In his paper 'Evolution, Causality, and Human Freedom. The Open Society from a Biological Point of view', **Franz M. Wuketits** points out that despite of the biogenetic heritage or even because of its efficacy, a self-planning and self-regulating social development has to be expected. The background of Wuketits' argumentation is formed by the proof that evolutive processes constantly and inevitable tend to produce more and more complex levels of self-organization and self-reproduction, in which course it is more and more probable that self-controlling processes will succeed. As a support for his concept, Wuketits refers to evolutionary epistemological theory whose development he has participated in and which describes the human ability for the perception of the world as a process of adaptation with the tendency to self-control and the potential of innovation.

This serves as a foil to **Michael Schmid's** essay 'Collective Action and the Selection of Rules' which offers a link to the sociological theory of acting. Schmid shows that it is possible to reconstruct the basic assumptions of the theory of social acting by means of a model based on the theory of selection and therewith understand and analyse the central social processes of distribution and reproduction as processes of structural selection.

In his essay 'Learning and the Evolution of Social Systems. An Epigenetic Perspective', **Klaus Eder,** on the other hand, reminds us that social evolution has to be considered as a process of 'collective learning' whose course is no more than indirectly influenced by external or genetical selection. Eder explains this process of learning as the accumulative process of the acquisition of collectively binding norms that shows autopoietic trends in as much as it keeps joining to existing knowledge on normative principles or rules and besides follows a course of development which is directed by the internal logic of that knowledge of rules. Eder corroborates his theoretical arguments by reflecting on the current anthropologic discussion on the genesis of the state.

Also **Peter A. Corning's** paper on 'Evolution and Political Control' proceeds to the evolution of political systems and stresses the selective profits gained by synergetically closed coordinations of acting if they are organized in a hierarchical form and by developping centers of control and regulation. Corning extenses his theory to a ecologic one in showing how different forms of organization and structure cause each other and reciprocally exert selective pressure. In the sense of a logics of modelling considered in such a way, Corning is right to claim that he presents a synopsis of a general theory of politics.

Bernd Giesen's paper 'Media and Markets' turns to the question how the reciprocal acting of numerous actors can be referred to in collective situations allowing neither extensive communicative votings nor the collective solution of questions for competence. Giesen finds an answer to his problem in the existence of specific principles of acting which he calls 'media codes' and whose coordinating effect depends on qualities that can be paradigmatically studied by the economic medium of money. Giesen shows in details the selective consequences of such media and specifies their chances for reproduction accordingly to the ability to regulate acting by means of processes of exchange.

Richard Pieper's essay 'The self as a Parasite. A Sociological Criticism on Popper's Theory of Evolution' only at first sight seems to be philosophical. But, actually, the author wants to free Popper's theory on the evolution of intellectual products from its imperfections by reconstructing it as a synergetic and ecologic theory. Poppers world 3 of objective traditions and cultural objects by this means has to be comprehended as the result of a process of collective symbolization whose detailed evolutive qualities are worked out against some central suppositions of Popper. This points out that the evolution of cultural entities, too, can be discussed within an extensive evolutionary model that, of course, has to be arranged in a different way than Popper's theory of (cultural) evolution with its idealistic connotation is providing for.

In the totality, all of these contributions offer a more detailed conception for the integration of a social theoretical formulation of problems in the range of a genuin evolutions theoretical context. They point out that social phenomena, individual or collective acts, as well as rules of acting or collective symbols, form systems able for evolution, whose genesis and transformations can be explained and described by the very theoretical instruments that is offered by an extensive and dynamic theory of evolution of a modern style. The editors of this volume emphatically hope that this demonstration effaces the impression that social theory was restricted to derive only from its own traditional sources and instead of this plead for an interdisciplinary learning for not to loose contact to models of explanation proposed in the range of the general theory of evolution and thus being able to solve the questions from whose unsufficient treatment social scientists in a strange misappreciation of their problems had deduced the methodological and theoretical independence of their subjects for a long time.

PETER MEYER

BASIC STRUCTURES IN HUMAN ACTION:
ON THE RELEVANCE OF BIO-SOCIAL CATEGORIES FOR SOCIAL THEORY

I. THE PROBLEM

In its intellectual tradition sociology has used biological ideas both as
heuristic devices for its models of social action, and as a source for the
anthropological question of where the "primary activity" toward action
comes from. Claessens has indicated (1980, p. 129, 229) that Webers's
famous definition of sociology includes a sort of constraint concept which
induces persons to permanently "be active, do something". Therefore,
sociological explanation of human action in terms of intentions and
reasons does not exclude a concept of primary activity which employs
causes as explanatory terms (Luckmann, 1980, p. 39). More recently Hammond
has demonstrated that Durkheim's sociology uses biosociological categories
to a large extent, even though these tendencies were "hidden as he
struggled valiantly to establish sociology as an independent discipline"
(Hammond, 1983, p. 124). Certainly sociology has been well established as
an independent discipline yet biology and other fields have continuously
provided it with ideas and models. For a number of reasons the nature of
this relation may be expected to persist: biology and its subdisciplines
are the most general among the life sciences, and sociology as a science
of social action must cope with potential implications of these more
general views.

Since theories and concepts change in biology as in any other field of
research, a permanent challenge arises for sociology. In former decades
concepts of specific "drives" and "instincts", or homeostatic mechanisms
concepts which seem to be outdated as general explanatory categories in
the life sciences, served as analogies for different social sciences. The
present paper does not intend to discuss the reasons for this
obsolescence, rather it submits to social science students some more

1

M. Schmid and F. M. Wuketits (eds.), Evolutionary Theory in Social Science, 1–22.
© 1987 by D. Reidel Publishing Company.

recent concepts from such fields as evolutionary epistemology and
sociobiology. Recent developments in evolutionary epistemology (Wuketits
(ed.), 1984) offer new insights into the crucial importance of cognitive
structures in evolutionary processes, a factor which has been frequently
underrated in general evolutionary theory and sociobiology. Therefore some
reductionist elements of sociobiology are criticized here, while emphasis
is placed on the fertility of sociobiological concepts (mainly
evolutionary stable strategies) as applied in a naturalistic explanation
of sociobiological categories, **e.g.** human affectivity.

Hammond has shown in detail the central importance of a biological
concept of affectivity in Durkheim's studies of social differentiation.
This paper will further discuss if and how recent naturalistic approaches
may contribute to the understanding of these biological aspects of
affectivity and its role as a basic structure in social interaction.

With regard to the question of "primary activity" some aspects of
physicalistic interpretations processes are presented in order to
demonstrate that both order and activity are common to such processes.
Behavioural systems are founded, it is suggested, in a number of
structural properties of physical processes, but these systems
progressively generate unprecedented levels of structural complexity. It
is further suggested that human action systems continuously interrelate
two distinct levels of behavioral integration, (a) the biosocial and (b)
the psycho-cultural (Schneirla, 1951), with (a) representing
phylogenetically successful behavioural strategies, **e.g.** evolutionarily
stabilized strategies, and (b) arising out of the needs of man's changing
physical, social and cognitive environments. Furthermore it will be
pointed out how sociobiology and evolutionary epistemology may be used in
the explanation of some features of the biosocial level, thus suggesting a
link between explanations in terms of causes and explanations in terms of
reasons.

II. SOME PRECONDITIONS OF BEHAVIOURAL PATTERNS

The general frame of reference for biosociology as well as for other life
sciences is evolutionary theory. Sociological theories have always
embraced evolutionary thinking. Most 19th century theories on societal

progressions such as A. Comte's and H. Spencer's did so (Mandelbaum, 1971, p. 77), whereas 20th century sociological functionalism was perhaps more influenced by certain organological interpretations of social systems. In the present context a few propositions about evolutionary theories will have to suffice.

Darwin's theory is certainly prominent among evolutionary theories; moreover it avoids certain vitalistic fallacies and may be considered one of the most successful theories ever proposed (Simpson, 1974). Its universal applicability has been further corroborated by molecular genetics and population biology, and at the same time it seems compatible with modern physicalistic explanations of the world. According to Eigen and Winkler the development of ever more complex molecules can be understood in terms of mutation and selection or vice versa (1981, p. 79).

As Eigen and Winkler suggest (1981, p.19), a physicalistic explanation of "primary activity" could begin with a sort of big bang theory which accounts for enhanced motion and energy as well as a number of "mutant" molecules; selective forces and "self-organization" stabilize some of these mutants. Molecules are provided not only with typical configurations of elementary particles but also "behavioural" characteristics, i.e. elements are kept in characteristic motion by molecular forces: "Every chemical reaction is connected with characteristic temporal and spatial patterns" (Eigen and Winkler, 1981, p. 110). Modern physics penetrated the nature of irreversible processes with such concepts as "entropy" and "dissipative structures".

The following implications for the life sciences should be mentioned: (a) The present day physicalistic interpretation of matter provides decisive insights into the laws and principles regulating motions and "behavioural" patterns in the microcosmos. (b) As Prigogine puts it "entropy thus becomes an indicator of development ... a time arrow" (Prigogine and Stengers, 1981, p. 113, transl. P. M.). The importance of "time" as compared to classical mechanics is stressed in another quotation from the same authors: "The direction of time ... is a precondition for all forms of life" (Prigogine, Stengers, 1981, p. 268, transl. P. M.). (c) Organisms, as well as other forms of matter, seem to originate in contingency; once existent "they create the conditions necessary for their propagation, i.e. their 'niches'" (Prigogine and Stengers, 1981, p.188, transl. P. M.).

The last implication seems to emphasize that the Darwinian theory of selection ought to be completed by this above-mentioned self-organizing tendency of the evolutionary process. Thus the ideal of perfect predictability as held by classical mechanics and by a variety of positivist schools, **e.g.** Comte's[1] impact on sociology, should no longer be upheld. Nevertheless, as will be indicated below, a number of sociobiologists have of late propagated this very ideal. To conclude this point: while Darwinian theory stresses the decisive role of chance, the slightest variation of parameters suffices to bring about new forms through self-organization. Mutations and their selective stabilization cannot be entirely accidental, as R. Riedl (1980, p. 41) points out; these stabilizations should be understood as approximations of forms to the prevailing laws of nature. Evolutionary theory thus includes causal as well as finalistic elements (Wuketits, 1981, p. 44). Behavioural patterns in the microcosmos, as mentioned above under (a), suggest that matter and, even more so, biological structures include order and activity (Prigogine and Stengers, 1981, p. 140). Organismic behaviour is not primarily a "reactive but an active phenomenon", a view entirely in accordance with a biosociological perspective. This consequently suggests the adoption of Bertalanffy's "active-dynamic" philosophy of the organism (Bertalanffy, 1949). Prigogine's concept of "dissipative structures" furthers general understanding on how this primary activity may be transformed into coherent "behaviour": "each molecule behaves as if it had information on the general state of the system" (Prigogine and Stengers, 1981, p. 171, transl. P. M.), as illustrated by an embryo, for example. Biochemical details of morphogenesis are outlined in Eigen and Winkler, and in Prigogine. For the present context it should be kept in mind that the complex interaction of genetic systems and the physical world requires an intermediate system, the phenotypic. Bateson (1982, p. 220) has interpreted this interaction as one between two distinct stochastic systems, the genotypic and the phenotypic, where the latter system relates the former to the physical world. As will be outlined below, sociobiology explains away some of the complexities of this interaction.

A final remark should stress the decisive importance of the "time arrow" for the evolution of cognitive structures and organismic behaviours. Among its results, the punctuation of events and development of an "individual, self-organizing and constantly improving memory" (Leinfellner, 1983, p. 231, transl. P. M.) are worth mentioning. Once existent, the DNA memory provides all individual organisms with a set of perceptory thresholds which select relevant external events and make communication systems possible. As indicated, perception implies that between event e_1 and e_2 time passes, and the elapsed time punctuates the dependent event, which in turn is perceived as a threshold, the real object of perception (Bateson, 1982, p. 251). Without the direction of time, life could not exist; even the amoeba searching for food could not do so and without "knowing" the difference between past and future (Prigogine and Stengers, 1981, p. 268). A world of "meaning" and goal-directed action, **i.e.** the real object of sociology, would not exist without inbuilt "mechanical" thresholds, such as the tremor of the eye as well as some physiological mechanisms which bring about appetence behaviours and cerebral structures giving rise to a common notion of time in the human mind. Phylogenetic constitution puts all members of a species into a "participatory universe" (Prigogine). The peculiarity of man has the effect, however, that he does not merely take part in a species-specific set of behaviours but that he additionally belongs to a variety of "pseudo-species" (Erikson, 1966, p. 340). Some of the implications of this pseudo-speciation will be touched upon later. First, however, a few conclusions on the previous considerations will be presented.

III. TAKING PHENOTYPES SERIOUSLY: CRITICAL REMARKS ON SOCIOBIOLOGY

An adequate theory of the progression[2] of living beings must cover the two distinct stochastic systems, the genotypic and the phenotypic. The latter system, as Bateson (1982, p. 224) points out, includes the contingent elements providing for change, whereas the first system determines the range of potential genetic alternatives. Since there is no direct relationship between genetic material and the external world, selective processes necessarily requires phenotypic mediation.

The relative importance of phenotypes for the entire natural process cannot be adequately understood by reducing it to a mere epiphenomenon of genes, as some sociobiologists seem to suggest. Rather, phenotypes should be seen as information about relevant properties of the physical world which enable genetic populations to produce new generations of individuals fit for life in this particular world. Thus genes, with their strategy of maximal replication, can only achieve this "goal" if their corresponding phenotypic system provides them with adequate "knowledge" of the physical world. As evolutionary epistemologists (including Campbell, Popper, and Riedl) have pointed out, this is perhaps the most general reason for the development of cognitive properties in organisms[3]: even a very simple structure among these, chemiotaxis, includes an adaptation to the direction of time and thus provides organisms with a basis for developing information processing behaviours, i.e. phenotypic elements which render possible maximal replication of genes.

Maynard Smith and Leinfellner have demonstrated how the development of complex behavioural systems and their corresponding cognitive structures can be explained by modern games theory, mainly the concept of evolutionary stable strategies. This approach will be further discussed below.

Cognitive faculties are a decisive fitness maximizing quality of organisms. This quality in turn is part of the organisms' phenotypic make-up. Cognitive systems then seem to be founded on some very general processes in the physical world: (a) Physical events obey the order of molecular motions and macro-structure impacts. (b) The time arrow expressing this order of motions relates past events to those of the present and future. (c) Organisms dispose of sensory receptors enabling them to associate external events and behaviors within a "reaction range" (Dobzhansky). (d) Cognitive systems furnish sets of evolutionarily tested hypotheses on external events and expectations of future events. Behaviours are designed to extract energy and thereby prepare for reproduction. (e) Human actions including mutually accepted "meaning" imply the general direction of time, complex sensory and neurological mechanisms, a complex cognitive system and some "emergent" socio-cultural entities. In summary, cognitive systems of progressive complexity may be seen as manifestations of Schrödinger's "order on order" principle (Riedl, 1980, p. 32): organismic learning processes have their foundation in "molecular patterns of experience which genetic material has extracted from its world" (Riedl, 1980, p. 45, transl.P. M.).

Before discussing some possible implications that these elements of natural processes may have on sociology, a short look at sociobiology will inspect the theoretical fertility of this discipline for the explanation of human social behaviour.

The theoretical scope of sociobiology has been outlined by E. O. Wilson. This new discipline facilitates explanations for a vast range of organismic behaviours, from those of the simplest species to man's. Wilson and other sociobiologists see the main advantage of their approach to be its evolutionary orientation. This makes more general explanations of behaviour possible, as opposed to those in the social sciences, and eventually will lead to the integration of "sociology, the other social sciences as well as the humanities" into sociobiology (Wilson, 1975, p. 4). It will not be necessary here to present details from sociobiological theory. Rather a few critical comments will be made on the general outlines of sociobiology, especially on its handling of the relation between genotypes and phenotypes.

Sociobiologists hold different views on the question as to what the potential contribution of their discipline to the understanding of human behaviour might be. Wilson, Dawkins, Trivers, to name but a few, are prepared to make propositions on human social behaviour, whereas Maynard Smith (1978a, p. 31), is rather reluctant on this point. The vivid discussions in the social sciences concentrate, of course, on sociobiologists' claims of providing superior insights into the nature of man and the working of his social organization. Whatever their differences, most sociobiologists would subscribe to a number of theoretical preconditions: (a) the adoption of a neo-Darwinian evolutionary theory which is mainly selectionist; (b) the consideration of genes as the unit of selection, which leads consequently to an intrinsically individualist approach; (c) the consideration of "altruism" as a genetic investment strategy which enlarges fitness of kin groups. Durham (1976, p. 102) suggests a terse formulation of the underlying principle of "phenotypic cost": "that there is a biologically limited amount of time and energy available to each organism. It is believed that natural selection adjusts the genetic influences on behaviour so that this time and energy ... are spent in ways that tend to maximize the representation of a given individual's genes".

Among human social behaviours to be explained by sociobiology Wilson
names the "biogram", the incest taboo, hypergamy as a typical marriage
pattern, and sex-specific types of aggressive arousal. Some of these
behaviours will be looked at more closely below after a brief discussion
of some underlying assumptions. Wilson (1978, p. 34) advocates a
concentration of sociobiological theorizing on primitive societies[4],
since these are thought to be closer to the "natural" order of things. At
least two objections can be made. While social organization of these
societies is certainly less complex than modern types, a biological and
anthropological dividing line based on sociological categories seems
untenable. Why should one type be closer to "nature" than the other?
Related to this objection but more specific, the incest taboo may well be
universal in man (Bohannan, 1963), but cultural ideas on what is to be
considered incestuous vary enormously. In summary, sociobiological
propositions on human behaviour cover but a very general set of structures
which require comparative research on the empirical variety of human
societies. The question remains, however, as to what possible contribution
human sociobiology could make to a theory of the "total social phenomenon"
(M. Mauss). Before returning to this topic additional objections will
attempt to clarify certain epistemological problems of sociobiology.

A number of critics have pointed out that sociobiology's
materialistic-mechanistic bias (Meyer, 1982, p. 103) brings about
propositions on human behaviour which are rather similar to other
reductionist strategies, e.g. Skinner's behaviourism (White, 1983, p. 23).
Moreover, Searle demonstrates that Wilson and some other sociobiologists
use intentional categories without properly considering their far-reaching
epistemological implications. Searle (1978, p. 175) elaborates this
problem by indicating that not a single description of behaviour can do
without intentional categories. Searle and other critics have shown that a
social science based on the separation of the two stochastic systems in
organisms, systematically underrates the complexities of genotypic and
phenotypic interactions and leads to simplistic theories of behaviour,
especially in higher species. As has been indicated above, some of these
epistemological shortcomings may be avoided by applying game theory
(Maynard Smith, 1978a; Leinfellner, 1983).

Numerous applications of this approach have demonstrated how maximum reproduction in populations can be explained without resorting to such categories as consciousness or rationality. Within a population, contestants for reproduction adopt different "strategies", i.e. certain "specification(s) of what a contestant will do in every situation" (Maynard Smith, 1978b, p. 261). Among these, a particular strategy, say X, turns out to be evolutionarily stable (ESS) and is subsequently protected against any mutant strategy because X increases fitness. Within particular populations those descendants will prove fitter who play ESS since this behavior will keep "phenotypic costs" low and at the same time maximize genetic offspring. From a behavioral viewpoint evolution can be seen as a succession of species-specific ESS combinations, maintaining optima between populations and their niches. As Leinfellner (1983) has pointed out, evolution as a general process brings about improved solutions of problems with progressively increasing intelligence as a concomitant.

Contrary to some reductionist views among sociobiologists, cognitive elements are a corollary of evolution; hence behaviours of living systems may not be adequately understood without taking these elements into consideration. To sum up, cognitive faculties built upon such very general properties as the time arrow, the "order on order" principle, or important aspects of self-conscious and rational actions may be understood by resorting to these properties.

IV. SECONDARY TYPE EXPLANATIONS DO NOT EXPLAIN AWAY PRIMARY TYPE EXPLANATIONS

Sociobiology's potential contribution to the understanding of intentional, purposive actions is an approach which renounces these categories to some extent.

Sociobiologists stress the importance of an evolutionary outlook on all aspects of behavioural systems. Thus Maynard Smith (1978a, p. 23) points out that different types of explanations, e.g. functional explanations, may easily be transformed into evolutionary categories. Transformations of this sort seem to be quite advantageous in their combination of functional analyses with the "survival value" of various elements of behavioural systems. According to Barash (1978, p. 23),

explanations stated in terms of primary causality can be differentiated
from those stated in terms of secondary causality. These terms, being a
major topic in recent sociobiology debates, will not be elaborated on in
the present context. Their theoretical fertility does not seem dubious at
all, rather the significance of the respective propostions for an
integrative perspective will be inspected. An example might be helpful
here: the invention and social dispersion of tables.

Comparative studies could probe into various sets of meanings of this
object and take table manners as its social manifestations. Further
studies could investigate differentiation of roles, strata **etc.** with
reference to records of table manners. A secondary research strategy
applied to this topic would certainly encounter difficulty: it would
perhaps point out some physiological aspects of eating habits or it might
concentrate on analogies in human and animal behaviour. Whatever the
potential contribution of these efforts they would necessarily fail to
convey those social meanings as reconstructed by historians and cultural
sociologists.

The situation seems similar to Eddington's problem with his "two"
desks: the first is his familiar brown wooden desk; the same desk from
another perspective is colourless, consisting of "myriads of nuclei and
electrons in a vast vacuum" (Agassi,1977, p. 156). Agassi, in a discussion
of Eddington's problem, points out that the desk, which, from a
physicist's view, is a collection of nuclei, does not "explain away" the
observable desk. A similar conclusion may be applied to the previous
problem: a physiological explanation of the eating process certainly does
not replace the study of cultural meanings applied to this process and
their theoretical reconstruction by social scientists.

Relations between sociological and sociobiological propositions on man
may be interpreted analogously. Sociobiology may explain sex-specific
behaviours, parental investment patterns and patterns of aggressive
behaviour. What it does explain in terms of secondary causality is the
survival value of a particular behaviour which does not exclude the fact
that humans may be additionally motivated by learned mechanisms, **i.e.**
phenotypic elements. Sociobiological explanations certainly cannot replace
explanations in terms of intentions or sociocultural values. By asking why
these particular structures evolved and what their contribution to

individual and population survival is, however, a basic structure of human actions can nevertheless be depicted. Provided that the incest taboo and some gender-specific behaviours are common to all human populations, explanations in terms of evolutionarily stable strategies may well be an appropriate manner to demonstrate the reproductive advantages of these behaviours. It should be emphasized, however, that sociobiology can merely explain why these phenomena are common to all human populations, while anthropology and sociology probe into the layer of primary causality. The very important contribution of a human sociobiology would then account for the basic structure of affectivity, "which give(s) depth of meaning to social life" (Hammond, 1983, p. 127). Affectivity provides specific situations in individual life cycles with differential arousal thus (1) restricting investments of vital energy to particular situations, and (2) generating a basic structure for the cultural process of applying "meaning" to those situations.

As will be pointed out below this concept of affectivity and emotions may be instrumental in seeing as "much mechanics as possible even in what may have meaning" (Agassi, 1977, p. 156).

The suggested relation of primary and secondary causality type explanations seems to omit some fallacies, notably those of reductionism. Furthermore it advocates the systematic study of interactions between the spheres of body and mind, whereby the former sphere provides for certain states of physiological arousal in social situations.

As an integrative concept for such research Count's term "biogram" (Count, 1958) is suggested. It stands for the study of morphogenetic characteristics and their corresponding physiological states and their combined effects representing the species-specific "physiological drama". As indicated above the concept of ESS bestows an additional explanatory thrust. Since it delves into secondary causality, it thus makes primary social mechanisms, cultural institutions and certain universal behaviours accessible to evolutionary explanation. In higher animals, however, especially in man, the sociogram representing the entire behavioural system is dominated by "downward causation". The major effect is that such entities as ideas are just as real as cells and molecules.

Among the various implications of "downward causation" only two directly relevant to the present context will be mentioned here. With regard to individual behaviour affectivity seems to provide the link between the system of cognitions acquired during socialization and the system of evolutionarily selected solutions to certain biological problems. Ideas on love, as expressed in social rules, obviously change with the prevailing characteristics of social structure, whereas ideas on the relation of the sexes are imbued with particularly strong affective qualities because they touch upon the major biological problem, **i.e.** reproduction. While cultural ideals of love vary enormously, every human group has in fact invented such ideals, and the corresponding social situations are associated with affective-emotional qualities. The proposition of "downward causation" obviously does not exclude the combination of cultural ideals with differential arousal patterns in an individual's acts.

By their behaviour which is an aspect of their phenotype, animals can alter their environments and consequently the selective forces influencing them. The migration of an animal population might serve as an example (Agassi, 1977, p. 157). This fact emphasizes the generative, self-organizing tendencies in natural processes. For the human species the enlarged mental capabilities offer additional prospects for influencing ecological conditions. Very seldom, though, do the overall consequences of human actions correspond to their primary intentions. Far from being trivial these unintentional consequences of human action result in a progression of social macrostructures, institutions of all sorts which again select individual and group actions. Socio-cultural evolution emerges as a new level of sociality, which is nevertheless the object of selective forces, old and new.

V. BIOSOCIOLOGY: A LEVELS MODEL OF MAN

As indicated above sociologists have introduced, tacitly or not, concepts of primary dynamics into their theories. Agassi (1977, p. 29) has pointed out why ideas of primary dynamics are of the greatest importance for sociology as well as for other social sciences in spite of their being beyond sociological categories. Any science requires a "metaphysics" as "sets of regulative ideas for science". Sociologism would be considered a kind of metaphysics denying any effect of biological laws on social actions, thus rendering sociology "the embracing framework or the foundation of all social sciences" (Agassi, 1977, p. 40). Contrary to such a position it is suggested here that sociological theories on human action necessarily embrace a number of biological assumptions in addition to sociocultural variables, including a concept of primary dynamics. Discussions on such assumptions will further the understanding as to why the most different cultures developed very similar behavioural solutions for certain social problems. This point will be further elaborated after some additional epistemological considerations.

According to the concept of "downward causation" ideas or ideals, **i.e.** cognitions are just as real as cells and molecules. Certainly man learns the vast majoritiy of these cognitions but "no individual learning would have become possible without the previous existence of differentiated patterns of molecular experience" (Riedl, 1980, p. 45). It will be outlined below why the distinction between preconditions of cognitions and those cognitions acquired during an individual life span is not only of phylogenetical interest but may offer some insights into man's actual behavioural integration as well. With regard to this integration two distinct levels are suggested by analogy: (a) a biosocial and (b) a psycho-cultural level.[5]

Turning to the characteristics of the biosocial level it should be indicated that they are made up of a variety of elements, only a few of which can be touched upon here. As an integrative concept Count's biogram is proposed (Count, 1958). It stands for the study of morphological characteristics and their corresponding physiological states.

Emotions[6], it is clear, are to a large extent expressions of the sequential organization of the biogram. Certain sequences are charged with special qualities of the limited resources of vital energy. At the same time social evolution influences many aspects of these sequences, in particular by applying various meanings to them. Affectivity and emotions must both be mistaken for any sort of "instincts" then.

To be sure, cultural values must not in the least be thought of as mere expressions, epiphenomena, of biological processes. Rather these ideas and values present specific solutions to certain problems raised by the human biogram. Their particular content, **i.e.** the respective cognitions as well as normative propositions are "free inventions"

(Langer, 1971, p. 326). Epistemologically these inventions belong to an
emergent level, the psycho-cultural, with socio-cultural selective
mechanisms deciding which of these cultural inventions are to persist.
Though non-reducible, the human biogram, as well as some peculiarities of
the mental system, set up structural constraints for this level. A few
more details of this paradoxical relation will be dealt with below after
considering some characteristics of human mental systems.

The time arrow is a decisive precondition for the evolution of
perceptory systems: it may be inferred that man's ability to perceive
punctuations of external events is founded on this evolutionary
background. Patterns of molecular "behaviour" present a point of departure
for organisms with perceptory systems, these systems being at the same
time furnished with genetically coded information on phenomena relevant
for their survival. This information provides organisms with expectations
of recurrent events. Thus "organisms do not simply wait for the occurence
of stimuli but ... are actively searching for them" (Riedl, 1980, p. 49).

It is quite clear that temporal structures encoded into mental systems
are not only a precondition for the "fitness" of corresponding behaviours
but even more so for the evolution of complex social behaviour, based on
mutually shared meanings. As Luhmann has pointed out (1976, p. 198) the
meaning of social acts always points to a future behavioural sequence.
Similarily, Luckmann (1980, p. 101), following Husserl and Schütz,
underlines the impact of "temporality idealizations" on the integration of
everyday social behaviours. These idealizations present syntheses of three
dimensions of time, the objective, the biological and the subjective
(Claessens, 1980, p. 132), with the biological dimension and its
particular internal organization being the precondition for psychocultural
definitions of the other dimensions.

Man's unique position in evolution becomes manifest in the paradoxical
relation of the two levels. The psycho-cultural level presenting man's
potential for penetrating into the laws of nature furnishes human
populations with new forms and increasing quantities of energy.

Apparently this growth is the single most important causal factor for
variation in the size of groups, social differentiation and stratification
(Adams, 1975, p. 217). The biosocial level seems to relate these
variations of social structure to human affectivity which has been

interpreted as a set of evolutionarily stabilized behavioural strategies. Two examples, the incest taboo and the density problem, will further illustrate the nature of this relation.

VI. THE INCEST TABOO: A BIOSOCIOLOGICAL VIEW

The incest taboo has been a classical topic among sociologists, anthropologists and psychoanalysts. Whereas there is agreement as to the universality of the taboo, there seems to exist no generally accepted explanation for it and disagreement concentrates on the question as to why different human groups should consider different types of matings illegitimate (Alland, 1980, p. 368). Consequently anthropologists continue to argue over problems such as: Why do some societies consider marriage of "first cousins who are the offspring of mother's sister or father's brother" as incestuous but permit marriage "with cousins who are the offspring of mother's brother and/or father's sister"? The following discussion will not be concerned with these details but will concentrate on the quasi-universality of rules against sexual relations between close relatives, **e.g.** father or mother and their children.[7]

For a long time a majority of sociologists subscribed to an explanation which suggested that the function of the taboo for the definition of kinship was the cause for its evolution (Parsons, 1975, p. 62; Hallowell, 1969, p. 472). The idea gained additional support from psychoanalytic interpretations of certain crises in ontogeny, **e.g.** the oedipal crisis in the process of personality formation. The argument that a particular cultural institution has evolved because of its "function" for the development of society will however have to deal with epistemological criticisms pointing to the general shortcomings of teleological explanations. Its major weakness is the use of a "function" which by definition appears developmentally later as an explanation for its evolution at a prior stage. With this background it seems more promising to employ a secondary, or evolutionary, strategy to outline causes for the evolution of a particular institution and then explicate the advantages of this institution for the working of society in terms of primary causality.

Sociobiologists have applied evolutionary theory, in particular population genetics, to the explanation of incest avoidance. These studies emphasize the risks of incestuous matings: "Among 161 children born to Czechoslovakian women who had sexual relations with their fathers, brothers or sons, 15 were stillborn or died within the first year of life, and more than 40 per cent suffered from various physical and mental defects" (Wilson, 1978, p. 37). Matings between close relatives, **i.e.** incest in a narrow sense, are extremely bad investments of reproductive resources and consequently one should expect such behaviours to be disfavoured by the natural process. In fact most animals also seem to avoid incestuous matings (Barash, 1981, p. 250). Hence explanations in terms of secondary causality are well-founded; why then should a cultural taboo be necessary to prevent man from such deleterious behaviours?

From an evolutionary viewpoint the development of mechanisms totally preventing this type of matings would possibly have turned out to be very costly, especially in higher species, since they might have reduced general behavioural adaptability. Competition between behavioural strategies is a less risky way to discover optimal solutions for the time being.

These considerations indicate why instincts or inborn release mechanisms (IRM) are not appropriate behavioral solutions in the case of man. The potential reduction of behavioural plasticity and genetic mutability would be too high. Studies on socialization processes in Israeli kibbutzim point to the formation of mental mechanisms in humans which reduce the likelihood of incest and thereby bring about adherence to the cultural taboo. These studies conducted by Joseph Shepher show that "a sexual aversion automatically develops between persons who have lived together when one or all grew to the age of six" (Wilson, 1978, p. 36). Furthermore this aversion is not restricted to people of actual blood relationship. For a more thorough interpretation of these findings the two-levels model will be employed.

As indicated above evolution seems to favour behaviours which maximize the reproduction of organisms under the condition of a limited amount of time and energy available to each organism. Sociobiological theory can specify why and how some behavioural strategies are evolutionarily stabilized in certain ecology-species settings. In connection with human behaviours this sort of reasoning can demonstrate advantages of particular gene strategies in terms of reproduction. It explains, why corresponding phenotypic behaviours have evolved and why they are imbued with different degrees of emotional arousal. The biosocial level of human sociality can be thought of as a set of evolutionarily selected strategies which operate in man's behavioural system as a differential arousal pattern for social interaction. Any behaviour of a reproductive nature, such as sexual orientations or parent-child relations, but defense of a person's vital resources as well, will be vested with considerable amounts of vital energy which can bring about differential arousal.[8] Shepher's study sheds some light on the interrelation of the two levels.

While some very undifferentiated emotional needs exist in the unborn child, these needs, or rather their potential elicitors, seem to be narrowed down in ontogeny. The ontogenetic formation of an incest avoidance mechanism possibly begins rather early in life with the delimitation of a number of persons with whom sexual relations, in the adult phase, are not likely to develop. This demonstrates that human affectivity is a rather undifferentiated behavioural program, and that emotions are hence malleable (Ekman, 1979, p. 178), over a certain period at least; the social group is thus enabled to further limit undesirable social attitudes in the child. The entire socialization process and the internalization of cultural institutions serves this purpose and fills the psycho-cultural level.

VII. THE HUMAN BIOGRAM AND THE ROLE OF CULTURAL INSTITUTIONS

The human biogram represents a number of biosocial patterns already tested by evolution, of information of favourable solutions to biological problems. Considering man's huge mental capacity and his related ability to adjust to nearly any ecological system on this planet, this information must not be as specific as instincts or IRM. Hence a genetic system will have to carry very broad information so as to leave the way open for phenotypic adjustment to a variety of ecological and social situations. Shepher's study throws some light upon the interrelation of the two levels; the biosocial patterns initially based on genetic information and those completed by childhood learning processes. The desired behaviour,

e.g. incest avoidance, is the result of emotional pervasion of certain types of social relations and of social norms. Social norms and values, in general terms, being the major content of the psycho-cultural level, relate persons to the requirements of particular societies; at the same time their variability is the most general precondition for the distinct selective processes of socio-cultural evolution. The nature of the interrelation of the two levels may now be characterized as a permanent comparsion of two information systems, the biosocial which provides particular sequences of the biogram with emotional qualities, and the psycho-cultural, made up of ideas potentially generative of variations of social values, group size and other macrostructure concomitants. Since both levels are products of evolution, one should expect a mechanism which minimizes the structural strain between them. Nevertheless tensions do arise for a number of reasons. Before returning to this point an additional consideration on cultural institutions will be presented.

These institutions are certainly part of the psycho-cultural level; any tribal society that desires to prevent incest resorts to a number of ideas, mostly of a mythological or religious character, thus legitimizing the taboo. Very seldom, though, can these ideas be considered logical inferences from theories of genetic reproduction. In fact, for a very extended period in human history, theories on reproduction were rather crude in terms of scientific rationality. How then can the very existence of this particular cultural institution be explained?

In man, any extended sequence of social behaviour has both a biological foundation and psycho-cultural elements due to "downward causation". Cultural institutions, as a special case of these elements, must contribute to reproduction, **i.e.** they must have been brought about by general selective processes. As indicated above, biosocial patterns represent tested behavioural strategies. However, in man these strategies take the form of affectivity and emotions, remaining adjustable to a variety of social elicitors. It is suggested then that affectivity is the necessary link between biosocial patterns and psycho-cultural processes, relating the sphere of major variation, the psycho-cultural, by means of emotional arousal to the evolutionarily tested biosocial strategies. According to this proposition affectivity procures a sort of feedback system between the two levels, thereby reducing structural strain. On the

macrolevel of human groups cultural institutions may be attributed the
same role, **i.e.** the minimization of such strain. As pointed out before it
would not be wise from an evolutionary perspective to avoid strain
entirely since any evolutionary system must allow mutant behaviours. In
fact, despite its being taboo, incest does occasionally occur, and causes
a strain on cultural rules. In consequence of such a deviance the variety
of social sanctions will now be implemented by the people concerned, thus
punishing deviant behaviour and at the same time ratifying the cultural
institution.

Obviously the taboo is an important factor in evolution's maximal
reproduction strategy. It keeps mutant behaviours at a necessary minimum
while defending the evolutionarily stabilized strategy. Shepher's study
discloses the ontogenetic formation of the taboo and its emotional
concomitants, with the cultural institution serving as an additional
control through social interaction and, during maturity, through implanted
social expectations as well. The hypothesis suggests then that this
particular institution has evolved because of its contribution to
reproduction. Under the conditions of man's vast mental capacities and
corresponding degree of individual autonomy, cultural institutions and
social value systems are much more appropriate solutions than instincts or
inborn release mechanisms.

Hammond's interpretation of Durkheim's studies on the relation of
social density and social differentiation emphasizes the huge impact of a
biosociological concept of affectivity on these studies. Similarly he
suggests that human affectivity, as any other part of man's biogram, must
have a selective advantage. In situations of growing social density
"individuals will begin to manufacture differentiation when density
changes begin to strain their limited affective tools" (Hammond, 1983, p.
128). In this concept affectivity seems to operate again as a sort of
feedback mechanism between group size and individual investments of scarce
affective resources, bringing about social differentiation as an
unintentional concequence. This differentiating effect of affectivity is
pointed out in numerous studies on social cohesion. Thus Moreno shows in
his classical studies (Moreno, 1967, p. 153) that persons are provided
with a limited emotional extension capability which restricts emotional
relations to a limited number of persons. Even more clearly this

differentiating effect comes to light in Little's studies on variation of social relations in battle situations (Little, 1964). In these studies persons tend to concentrate interactions of a positive emotional character on just one other person. Under the conditions of extreme danger the limitation of affective-emotional exchange to a "buddy relation" is obviously the optimum behavioral strategy. While social interaction splits up formal groups into dyadic cliques these units enable those persons involved to cope with some of the difficulties imposed upon them.

Living in social groups certainly belongs to the universal characteristics of the **animal sociale**, however these groups have varied enormously in form and size throughout social evolution. The studies referred to above point to a general emotional preference for comparatively small groups and it is this preference which brings about the major features of social differentiation. Perhaps sociobiology offers, for the time being, the best explanation for the evolution of this particular element of affectivity: Kin selection reserves mutual aid to immediate kin and brings about higher fitness for group members (Barash, 1980, p. 113, 257; Markl, 1982, p. 37). Human affectivity transmits this evolutionarily tested information on fitness-maximizing limitations of group size and applies it by analogy to the non-kin groups of more recent societal stages. While affectivity persists as a biosocial structure through social evolution, some of its emotional expressions, change as cultural meanings change, while other emotional expressions, e. g. facial expressions, seem to "universally signify particular emotions" (Ekman, 1979, p. 176). Research is under way in a number of disciplines including human ethnology and psychology as well as sociology to determine the relative importance of biosocial patterns in emotional expressions. For the present context it is a major goal to connect the various contributions towards a biosociological theory of human affectivity and outline this theory's relevance to sociological theory in general.

VIII. CONCLUSION

In years to come it is hoped that biosociology will extend its explanatory powers by completing the list of biosocial elements in human behaviour and continuing to integrate the findings from various disciplines. A number of new disciplines ranging from ethnology and sociobiology to evolutionary epistemology have provided new insights into the biological bases of human behaviours. While some of these disciplines definitely follow reductionist strategies, it has been suggested that biosociology must consider the applicability of these strategies very critically. As a science of man, biosociology should clearly follow an anti-reductionist route without totally denying, however, the fertility of reductionist explanations for limited purposes. Sociobiology presents itself as an example for critical analysis: Since most of its methodology is reductionist it nevertheless offers insights into the foundations in evolutionary economy of a number of man's social behaviours. As indicated above some of the shortcomings of these reductionist explanations may be countervailed by resorting to a sort of gamestheoretical approach as propagated by Maynard Smith and Leinfellner. Hence sociobiology's genetical reductionism is to be supplemented by considerations from evolutionary epistemology and other perspectives whereby the role of "downward causation" and cognitions in integrating organisms' behaviours may be more adequately understood.

The relevance of biosociological considerations for general theories of sociology does not need special emphasis after all: Biosociological assumptions are in fact a necessary element of any general sociological theory. A major future task of biosociology should be the further clarification of its concepts by examining findings from other life sciences and subsequently introducing these concepts into discussions of some of sociology's most fundamental premises.

NOTES

1. The ideal of perfect predictability is implicit in Comte's central idea "that there was a necessary progression from stage to stage in the evolution of mankind" (Mandelbaum, 1974, p. 67).

2. For a differentiation of "progression" and "progress" see Simpson (1974).

3. A general introduction to the philosophical background is provided by Popper and Eccles (1977); see more specifically on "philosophy of the organism" Sperry (1965).

4. The term "primitive" is used as by Douglas (1978, p. 77): "Thus primitive means undifferentiated; modern means differentiated".

5. Ideas on this model were influenced by Schneirla (1951).

6. Human affectivity may be understood "as the physiological capacity to generate emotions" (Hammond, 1983, p. 125).

7.   Incestuous matings in this restricted sense have actually been reported from the Pharaohs and Incas and a number of tribal societies. In these cases incest was reserved to members of the higher social strata and royal families. A major function of corresponding social rules may be seen in the emphasis of the special status of these persons. So in a strict sense, the term "quasi-universality" is appropriate here.

8. Zimmer (1981) outlines a "logic of emotions", i.e. the survival value of emotions is discussed.

MICHAEL RUSE

EVOLUTIONARY MODELS AND SOCIAL THEORY.
PROSPECTS AND PROBLEMS

I. INTRODUCTION

Charles Darwin's most important work, **On the Origin of Species by Means of
Natural Selection**, was published in 1859. In that volume, Darwin argued
that all organisms are descended by a slow, gradual, natural process of
transformation - that is to say "evolution" - from one or a few original
forms. Additionally, Darwin proposed a mechanism for this process which he
called "natural selection". Darwin hypothesized that many more organisms
are being born than can possibly survive and reproduce, and that
consequently there will be an ensuing "struggle for existence". More
precisely, there will be a "struggle for reproduction". Noting that
organisms seem to differ between themselves, Darwin argued that success
in the struggle will be in part a function of the distinctive
characteristics possessed by the successful, and this he maintained would
lead to a natural form of winnowing, or selection. Given enough time,
full-blooded evolution results (Ruse, 1979).

Darwin said little in the **Origin** about our own species, **Homo sapiens**.
However, what everyone realized was that if indeed evolution through
selection be true, then we humans can no longer regard ourselves as the
favorite creation of Almighty God on the Sixth Day. We are, rather, merely
modified monkeys. (No one, including Darwin, has ever claimed that we are
descended from today's monkeys or apes, but rather that we all have a
fairly recent common ancestor.) Because of the shocking implications,
evolutionism was at once regarded as more than merely a scientific theory,
being instead a world picture with implications reaching into all areas of
our thought and action, including our moral thought and action. Darwinism
in various forms was taken to be an aid (or concersely, a threat) to

23

M. Schmid and F. M. Wuketits (eds.), Evolutionary Theory in Social Science, 23–47.

proper thinking about human behaviour, individually and in groups, and
such a view has persisted until this day (Ruse, 1976b).

In this discussion, I want to look at some of the most significant
inter-relations between our thinking about evolution and our thinking
about the whole broad area of social theory. I shall begin in the past
with a brief hisorical prelude, and then move to the central focus of my
interest, the present.

II. SOCIAL DARWINISM

As a matter of historical fact, the relationship between evolutionary
theorizing and social theory pre-dates the publication of Darwin's **Origin**,
for indeed, Darwin himself was much influenced by certain social doctrines
in the very discovery and elaboration of his theory of descent with
modification (Young, 1969). As is well known, Darwin became an
evolutionist towards the end of or shortly after the time he spent touring
the world on **H.M.S. Beagle**. Nevertheless, from the beginning, Darwin
realized that evolutionism in itself was not enough. It was necessary also
for him to find a mechanism of evolutionary change. For various reasons,
primarily because he himself came from a rural background and was much
aware of the success that early nineteenth century agriculturalists were
having in improving livestock and crops, Darwin soon realized that the key
to change lies in some form of selective process - one picks out the
organisms that one desires and breeds from them. However, for several
months (in 1837 and 1838), Darwin was puzzled as to what force might lie
behind any natural kind of selection.

Finally, in September, 1838, Charles Darwin read the major work on
political economy by the clergyman, Thomas Robert Malthus, **An Essay on the
Principle of Population** (1826). There, Darwin learned of the struggle for
existence, and at once he had the key insight necessary to produce a
natural form of development. The struggle which is ongoing in the world
ensures that only a few organisms survive and reproduce, and that those
which do tend to have characteristics distinctive from those which do not.
Given enough time, this leads to full-blown change. Moreover, it should be
noted that this is not simply change, but change of a particular kind.
Natural selection brought about by the struggle for existence leads to

change of an adaptive nature. Organisms are not simply thrown together as it were. They are, rather, integrated - functioning in a harmonious way - towards their own ends. They are adapted (Ospovat, 1981).

What sort of work was this volume of Malthus that so influenced Darwin? It was, in fact, a social tract of the most reactionary and conservative kind. At the end of the eighteenth century, appalled by what he believed were unreasonable hopes of progress, Malthus had argued that in the human species there is necessarily an ongoing struggle for existence. Human food supplies can be increased at best at only an arithmetic rate, whereas population numbers have always a potential for geometric increase. Since geometric increase outstrips arithemetic increase very rapidly, Malthus concluded that therefore there must be some sort of crunch. Using this conclusion, Malthus went on to claim that progress is impossible, and moreover that state-supported welfare schemes are misleading and dangerous. At best, one can put off disaster for one generation (Flew, 1963).

In later versions of his essay, Malthus somewhat modified his positions. Not a few pointed out that his doctrine was, to say the least, somewhat incongruous coming from the pen of a clergyman of the Church of England. Malthus therefore argued that the struggle could possibly be abated or avoided by what he termed "prudential restraint". However, Malthus was not sanguine about the possibilities. Consequently, he continued to argue for a view of society which still finds favour in certain reactionary quarters even today.

Darwin took Malthus' social theorizing and turned it on its head! He generalized the struggle for existence from the human world to the world of animals and plants, and agreed with Malthus that a struggle for existence is an inevitable phenomenon. However, having done this, far from using it to deny the possibility of any ongoing change, Darwin used it to make the struggle the mainspring of ongoing change. Darwin, as we have just seen, put the struggle behind natural selection, and this therefore leads to full evolutionary development.

Nevertheless, Darwin himself was always careful to dissociate his theory from any social implications. In particular, what Darwin certainly did not want to do was resuscitate the progress - moral or otherwise, real or apparent - against which Malthus fulminated. Darwin was insistent

always that evolution is a blind, directionless process. Of course, down
through the generations, complexity has evolved since the earliest, most
primitive forms, but Darwin felt that this change is a purely contingent
matter. It was here that Darwin felt, with some justification, that he had
made his most decisive break with the past. Although, like the Christians,
Darwin took the notion of adapation and functioning very seriously, for
him it was always a relative process. The eye that we have is something
which functions better than eyes of our predecessors, or, in a more
absolute sense, over the inadequacies of those who are born totally blind.
Nevertheless, Darwin was insistent always that adaptations like the eye
are not perfect, as one might expect had a Good Designer stood behind
creation, and moreover, Darwin was clear in his own mind that the ongoing
prosess of evolution is without any ultimate direction (Ruse, 1986).

Hence, although Darwin drew on Malthus for his theory, he tried to
keep his scientific thinking as clear as possible from political or other
moral implications or prescriptons. We cannot see in evolution any
justification for the moral worth of evolutionary change, nor for the
cherishing of the causal processes of evolution. This is not to say that
Darwin, as a very successful member of the Victorian middle classes did
not have some fairly conventional social views which frequently get into
his writings, particularly in the **Descent of Man** (1871). Darwin was a
grandson of the industrialist Josiah Wedgwood and expectedly had much
sympathy for capitalism. Nevertheless, as one reads Darwin one sees that
almost always he is trying to keep his science and his politics apart.

However, although Darwin wanted to keep his scientific theorizing
separate from his social beliefs, others were far less particular. Indeed,
there were many in the mid and late Victorian period who positively
welcomed Darwinism as the foundation for a whole new system of ethics with
profound and far-reaching social implications. These were people who saw
with the coming of evolutionism the death of conventional Christianity and
with it the failure of conventional supports of morality (most
particularly, the decline of the obligation to follow God's Will). They
therefore turned to evolutionism, as they variously interpreted it, and
sought guidance there for moral insight.

Most notable amongst these "evolutionary ethicists" was the all-round
thinker, Herbert Spencer (1857, 1892). Nothing loathe, Spencer was more

than happy to take Darwinism, or at least his version of it, and push it to the limit, deriving or devising a full-blown moral system. As is well known, Spencer argued for a fairly extreme laissez-faire socioeconomic system, where the government is supposed to take a rigid, hands-off approach to social problems. It is argued that unbridled capitalism is the right and proper, not to say most efficient, social system. This, at least was the theme of many of Spencer's supporters and admirers. For instance, we find the American theorist, William Graham Sumner, writing as follows: "Let it be understood that we cannot go outside of this alternative: liberty, inequality, survival of the fittest: not liberty, equality, survival of the unfittest. The former carries society forward and favours all its best members; the latter carries society downwards and favours all its worst members" (Sumner, 1918; see also Russett, 1976).

How did Spencer and his followers support this doctrine which became known as "Social Darwinism"? They invoked the very idea which Darwin had rejected and indeed at times explicitly argued against. They found within the evolutionary process a progressive move upwards and onwards. They thought that evolution shows real progress or advance from amoeba to man as it were. Thus, they argued that any help one can give to the evolutionary process is in itself thereby of inherent value (Spencer, 1857). Since the evolutionary process proceeds through a struggle for existence, the Social Darwinians claimed that all backing for such a struggle at once gains inherent value. Since that social system which seems to give most support for the struggle is one of laissez-faire capitalism, the Social Darwinians concluded that therefore such a sociopolitical approach to life is morally demanded.

Spencer and his followers had their critics, most notably "Darwin's bulldog", Thomas Henry Huxley (1893), who argued that morality lies in opposing the struggle for existence rather than aiding and abetting it, and such critics have continued down to this day. Indeed, I trust it is no longer necessary for us to dwell long on Social Darwinism for its flaws are readily seen by all. Most particularly, whatever one might think about the virtues and ills of capitalism, it is simply not the case that evolutionism gives support to the progressionism on which Social Darwinism rests and relies (Williams, 1966). As Darwin presciently noted, the evolutionary process is simply going nowhere, and rather slowly at that.

Thus, even if natural selection did lead to all-out bloody warfare,
unrestrained selfishness is alive at the heart of capitalism (and as we
will see shortly, there are serious questions as to whether it does have
these implications). There is no case for moving directly from the
biological state of affairs to the socially desirable ends. Without
progress, traditional evolutionary ethics, including Social Darwinism,
collapses.

III. ANIMAL SOCIOBIOLOGY

To most thinkers of the past hundred years, this collapse has ended
matters. The barriers between biological theorizing and social theorizing
have seemed impassable, and there has been little inclination to go over,
under or around them. Biologists have been absorbed in their own ideas,
and social theorists and others (like philosophers) have felt that there
was something fundamentally wrong about any attempt to derive social
insights from scientific hypotheses. Usally, the name of the eighteenth
century philosopher, David Hume (1978), has been invoked, and we are
reminded of his famous distinction between claims about matters of fact
and claims about values (the "is/ought" dichotomy), and it is pointed out
that all evolutionary ethicizing clearly violates this distinction
(Flew, 1967).

However, in recent years, a number of thinkers, particularly
biologists, have been arguing that perhaps matters are not so definitively
settled as was once concluded. The major reason for this change of mind
has been the growth and development of a sub-branch of that version of
Darwinian evolutionary theory which is the dominant paradigm in biological
studies today. This sub-branch is that which deals with behaviour,
particularly social behaviour between animals. In this and the next
section, I shall deal with the basic science. Then, in the following
sections I shall go on to consider the supposed implications for social
theory.

Darwin himself, in the **Origin**, was much interested in the problem of
social behaviour, and he wrote at length on questions to do with
instinct, particularly amongst the social insects, the ants, the bees, and
the wasps (Ruse, 1980). However, this sub-area of evolutionary studies

lagged behind other areas like systematics and genetics. There are fairly obvious reasons for this. It is far easier to study the eye colour of a fruit fly than its mating behaviour, particularly given that the latter is often distorted by experimental conditions and other like factors. Also, one suspects that with the growth of the social sciences in this century and their strong and natural desire for autonomy, there was somewhat of a counter-force against biological theorizing about behaviour. However, in the past twenty years or so, things have changed very drastically indeed. The study of animal social behaviour has turned from being virtually an non-subject to being one of the hottest areas within the evolutionary family, the kind of subject which attracts all sorts of bright young students. It has indeed now even received a new name, namely, "Sociobiology". Many works and papers have been written on the subject. Probably the best introduction is that of Harvard biologist, Edward O. Wilson, **Sociobiology: The New Synthesis** (1975; see also Ruse, 1984a).

Basically, the major theme of the sociobiologists is that the interactions which take place between animals, particularly interactions within the same species, are not necessarily of a kind which lead inevitably to a bloody struggle for existence. This conclusion is drawn, despite the fact that sociobiologists are no less committed than Darwin to the overall importance of natural selection as the cause of biological change, and indeed that like Darwin, the socio-biologists are strongly committed to the belief that selection must always act ultimately for the benefit of the individual rather than the group. Fairly obviously, were one to claim that natural selection benefits group interactions, then social behaviour would be a natural and expected consequence. (This group-favouring perspective has in fact been argued, most recently by the ethnologists like Konrad Lorenz (1966), and indeed has a long tradition going back to the work of natural selection's co-discoverer, Alfred Russel Wallace (1870).)

How do sociobiologists explain social behaviour from an individual selection perspective, that is, how do they account for what they term "altruism"? A number of mechanisms are proposed, but essentially the idea is one of enlightened self-interest (Dawkins, 1976). It is argued that cooperation frequently brings greater returns to the individual, that is to say greater biological returns, than all-out conflict and hostility. Two mechanisms in particular are proposed. Let us look at them briefly.

The first putative mechanism is so-called "kin selection". Today, it is realized, thanks to modern genetics, that what counts in the evolutionary perspective is not simply surviving, but reproducing. But reproducing means passing on copies of ones units of heredity, namely the genes. In other words, the key to biological success is transmitting one's genes. But remember that we all have genes, and that we all share genes whith other people, namely close relatives. For instance, I have half of my father's geneses, my son and doughter will likewise have half of my genes, and I share a fifty per cent similarity in genetic makeup with my siblings. Other relations are somewhat less-closely related to me, but there is still some relationship, even as one moves away from me. Now, since I share copies of my genes with close relatives, this means that inasmuch as they reproduce, I likewise reproduce, vicariously as it were. When, for instance, one of my brothers has children, they receive copies of his genes, but since my genes are like my brother's genes, they receive copies of my genes. What this means in effect is that instead of simply reproducing directly myself, my biological ends are furthered when my close relatives reproduce. Therefore, any help or aid that I can give close relatives in reproduction rebounds biologically to my benefit.

This is the idea behind kin selection (Maynard Smith, 1978). Normally, of course, one is better off doing one's own reproducing, since except in the case of identical twins, one is more closely related to oneself that to others, and therefore one best furthers one's biological ends by doing one's own reproduction, but, other factors can intervene. Indeed, William Hamilton (1964a, 1964b) showed in one of the most brilliant biological strokes of this century that in kin selection lies the secret to hymenopteran sterility. Why is it that ants and bees and wasps have sterile castes of workers who selflessly give their lives to the nest? As Hamilton pointed out, the answer is that because of a rather strange sexual system, female siblings are more closely related to each other than are mothers and daughters. This means that in the hymenopteran world one can better one's reproductive interests if one is a female by raising the daughters of one's mother than by raising one's own daughters. Consequently, hymenopteran sterility is highly altruistic in the sense of involving a great deal of cooperation, and yet it is at the same time activity which rebounds to the interests of the actor rather than to the group as a whole. Just what Darwinians expect and presuppose.

The other major mechanism proposed by sociobiologists to account for social behaviour is so called "reciprocal altruism" (Trivers, 1971). It is pointed out, quite simply, that there are times when one needs help, and the best way of getting such help is by being prepared to offer help in return. The idea is very much like an insurance policy. I have no particular desire to pay my hundred dollar premium. However, I realize that if I do pay my hundred dollar premium, then if I am in need, I shall be able to draw on, say, a ten thousand dollar payment. The same holds for others. The returns, or at least the possibilities of returns, are very much greater than the cost to myself. Although I lose out on my hundred dollars, this is a mere inconvenience, whereas were my house to burn down completely, I should be utterly ruined. The same idea, it is argued by sociobiologists, holds throughout the animal world. Animals aid others, including non-relatives, because there are the possibilities of reciprocating benefits. Thus, altruism evolves once again not because of some group or other supposed benefit, but because of returns or potentiality of returns to the individual actor. (A good review of animal sociobiology, theory and evidence, is given by Barash, 1982.)

IV. HUMAN SOCIOBIOLOGY

Had sociobiologists kept their theorizing and empirical investigations down at the level of the animals, then I have little doubt that they would have earned praise all round. It is, as I have implied, not merely an extension, but a very exciting extension of ideas implicit within the concept of evolution through natural selection. However, sociobiologists like Edward O. Wilson have argued at once that their views apply not merely to the brutes, but also to ourselves. They want to argue that all our behaviour, including our social cooperative behaviour, rests ultimately in the natural selection of favourable genes. Consequently, everything must be understood as in some sense rebounding to the individual. (See, especially, Wilson, 1978.)

This extension of their theorizing to the human realm has brought upon the sociobiologists extreme criticsm within and without biology. Much has been written about this controversy, and since I have myself discussed it elsewhere, I shall only touch on it in this paper (Ruse, 1984a; Lewontin

et al. 1984). It is argued that sociobiology is bad science, that it relies on inadequate forms of explanation, that it is not testable, and that even if it were testable proper testing is not being carried out. Analogously, it is argued that sociobiology is simply Social Darwinism dressed up in modern guise, and carries within it all the vile reactionary social implications that were so cherished by successful capitalists in the nineteenth century.

I believe that the simple answer to charges such as these is that it was indeed true that some of the earlier sociobiological writings were nothing like as careful, methodologically or morally, as they might have been. Nevertheless, ultimately, such charges miss the mark. It is true that sociobiologists argue that ultimately all benefits must relate biologically to the individual; but, to suppose that this implies that all individuals must be explicitly and literally selfish is simply to confuse cause and effect. What is argued by sociobiologists is that, down at the level of biological cause, returns must be made. But, no one thinks that this means that each individual must consciously be aware of (let alone, aiming for) biological reward. Obviously, such a view would simply be ludicrous in the world of hymenoptera. No one thinks that ants are conscious beings, or deliberately planning their personal ends. No more is it argued that such awareness must necessarily be present in the human world. In fact, sociobiologists argue that we usually all function a great deal more efficiently, if we are unaware of our biological ends, rather than if we are consciously trying to put such ends into effect (Trivers, 1971; Alexander, 1979).

Under normal circumstances one would try as far as possible to keep motives out of the picture, but since the critics of sociobiology have accused its practitioners of being little more than apologists for the capitalist **status quo**, one can legitimately point out that many of the critics have themselves hardly been disinterested enquirers. On the one hand, one finds many critics come from the social sciences, and show in word and action that they feel very much threatened by the possible success of sociobiology (Geertz, 1980; Sahlins, 1976). On the other hand, those from within the biological community who have been most opposed to sociobiology have often implicitly, and indeed at times quite explicitly, made clear their commitment to alternative ideologies, particularly

Marxism (Lewontin **et al.** 1984; Gould, 1978, 1979). Such critics feel strongly that human nature must be understood as an effect of social forces rather than as something relating to our evolutionary past. Hence, the vigour with which sociobiology has been opposed.

I do not claim here that sociobiology, and I include especially the human variety, is a perfect or completed science. However, I reiterate that there is not yet good reason simply to dismiss it as fascism by another name.

V. THE EVOLUTION OF MORALITY

Let us leave the controversy behind. What does human sociobiology have to do with social theory? Quite simply, it is the claim of the sociobiologists that their subject gives us a true and deep insight into the evolution of morality, and that once such insight has been achieved, much is made clear about the nature and status of such morality (Ruse and Wilson, 1986). There are two claims here, one scientific, and the other more philosophical. Let us take them in turn.

To get at the evolution of morality, we must look first at the most recent thinking about sociobiological nature of human beings. This has come in works that have been authored by Edward O. Wilson, and a young physicist, Charles Lumsden. They argue that the human mind is not a **tabula rasa**. It is, rather, an entity which is defined and constrained by certain innate principles (Lumsden and Wilson, 1981; 1982). These principles, or propensities, Wilson and Lumsden term "epigenetic rules". More formally, they write of epigenetic rules:

> Any regularity during epigenesis that channels the development of an anatomical, physiological, cognitive, or behavioral trait in a particular direction. Epigenetic rules are ultimately genetic in basis, in the sense that their particular nature depends on the DNA developmental blueprint. They occur at all stages of development, from protein assembly through the complex events of organ construction to learning. Some epigenetic rules are inflexible, with the final phenotype being buffered from all but the most drastic environmental changes. Others permit a flexible response to the environment; yet even these may be invariant, in the sense that each possible response in the array is matched to one environmental cue or a set of cues through the operation of special control mechanisms. (Lumsden and Wilson, 1981, p. 370)

Lumsden and Wilson distinguish between primary epigenetic rules, which
deal with our perception of the physical world, and secondary epigenetic
rules, which lie within the mind as it were, waiting for the raw
information from outside, which can then be controlled, manipulated, and
used towards human ends. A prime exemple of a secondary epigenetic rule is
our propensity to avoid incestuous relationships. **Prima facie** there is no
reason why one should not be sexually attracted towards close relatives.
However, there are very good biological reasons why this would be
deleterious, namely that close inbreeding leads to horrific effects.
Consequently, argue the sociobiologists, there are good biological reasons
why we should have a propensity to avoid incestuous relationships, and it
is the thesis of Lumsden and Wilson that the way in which this avoidance
is mediated is through a secondary epigenetic rule. We simply become
aware, innately, of a disinclination to breed with close relatives, and a
feeling of repugnace.

The epigenetic rules are of course only innate to the individual.
There is no claim that they represent ideas in the mind of God or some
such thing - the kind of innate ideas against which John Locke argued in
his **Essay on Human Understanding** (1975). Rather, it is claimed by the
sociobiologists that the epigenetic rules have come about through normal
evolutionary processes, and thus have adaptive value to us humans. Putting
the matter simply, those of our Australoplithecine would be ancestors who
were disinclined to mate with close relatives succeeded better in the
struggle for survival and reproduction than those who found their sexual
delights close at home.

The language of epigenetic rules leads naturally to an understanding
of the evolution of morality (Ruse, 1985). There is no logical or absolute
biological reason why humans should be moral. We might, for instance,
simply have been programmed to act blindly, like the ants. However, this
would have carried with it great disadvantages, namely an inability to
react flexibly in the case of changed situations. We might, conversely,
have been given super-minds which would enable us to calculate our best
biological interests whenever faced with a situation. There is nothing
moral (or immoral) in buying insurance. One calculates one's best
interests and proceeds forthwith. It could quite possibly have been the
case that we have no moral sense whatsoever, but simply calculated our own

best interests whenever a conflict or social situation arose. However, argue Lumsden and Wilson, this purely rational faculty would likewise have carried disadvantages, namely the need for humans to have very large and complex reasoning abilities, not to mention the fact that such a process of personal calculation would require time, which might not always be available.

Therefore, conclude the sociobiologists, morality is a compromising short-cut, between no reason and total reason. We are given a certain sense of moral awareness that we ought to act kindly and altruistically (in the literal sense) towards our fellow humans. This gets us to interact socially, which as it so happens, (unbeknownst to us), furthers our reproductive interests. In other words, the claim by the sociobiologists is that altruism in the literal sense of being nice to others leads to "altruism" in the metaphorical sense of furthering our own reproductive ends through social interactions.

What form are moral insights supposed to take, given the sociobiological hypothesis of foundation within the epigenetic rules? Here the sociobiologists refer to fairly standard and conventional work by moral and social theorists. After all, where else would they turn? It is argued that the epigenetic rules give us a sense of moral and social obligation very much along the lines of that proposed by philosophers and other enquirers into the nature of morality. Consider, for instance, one of the deservedly best-known attempts in recent years at formulating an adequate theory of justice, namely that of the Harvard philosopher, John Rawls (1971). Rawls argues that what is right and just for each individual is what would turn out to be fair.

> The guiding idea is that the principles of justice for the basic structure of society are ... the principles that free and rational persons concerned to further their own interests would accept in an initial position of equality as defining the fundamental terms of their association. These principles are to regulate all further agreements; they specify the kinds of social cooperation that can be entered into and the forms of government that can be established. This way of regarding the principles of justice I shall call justice as fairness. (Rawls, 1971, p. 11)

Elaborating, Rawls argues that in order to find the specific principles of justice, it is necessary for us as it were to put ourselves behind a veil of ignorance. We must make the assumption that we do not know where we will be in society, whether we will be rich or poor, clever or rather

dull, healthy or handicapped, or whatever. Obviously, argues Rawls, if we
knew our position in society and knew, for instance, that we were going to
be very healthy and born to a good rich family, then our natural
inclinations would be to maximize our own self-interests, and to argue
that healthy rich people ought to be favoured above all others.
Unfortunately, if we were born poor and handicapped, then given such a
favouring of the fortunate, we would be at a greater loss because of our
social situation than even our initial physical calamities put us at.
Therefore, argues Rawls, arguing from behind a veil of ignorance - not
knowing what our future position will be - what we will do in order to
maximize our own interests will be to devise principles of justice or
fairness. A just society will be one which will do the most for everyone,
given that they do not know what their position within that society will
be.

I should add parenthetically that Rawls is not preaching some form of
Utopian socialism, as many have assumed. He certainly allows for
inequalities within society; but, given his first principles, these
inequalities must be such that everyone would benefit more than were there
no such inequalities. In particular, for instance, one could argue that
doctors should be paid more than others, but, the argument for this is not
because one knows that one will oneself be a doctor. Rather it must be
because of a presumption that the biggest attraction to medicine is money
and that the only way of luring the most able students into medicine is to
pay them more. Since we all obviously benefit by better-quality doctors,
it is worth paying doctors more than the rest of us. This conclusion is
compatible with the enlightened self-interest that emerges from decisions
behind the veil of ignorance.

I need hardly say how readily this kind of analysis of justice within
a society lends itself to a sociobiological underpinning of the kind which
has been sketched earlier in this section. At all levels, we have self-
interest operation. This is not to say that in a Rawlsian just society
people are consciously acting in their own interests any more than they
are in a sociobiologically formed society. It is rather that it is self-
interest of some form which is the original causal factor - a factor of
which the participants might well be in a state of ignorance. The point is
simply that it is a natural move to argue that our sense of morality, as

embedded within the epigenetic rules, is that very sense that has been so brilliantly explicated and developed by philosophers like Rawls.

In other words, the sociobiological approach to social theory does not in principle call for radical revisions of our social thinking, nor, most particularly, does it call for a return to Spencerian Social Darwinism. Rather, subject to some provisos to be noted later, the sociobiology of humans confirms and elaborates social insights which have been independently articulated by some of the foremost, most forward-thinking social theorists of our day.

VI. THE STATUS OF MORALITY

This now brings us to the second, more philosophical question, namely, what, if anything, does this new evolutionary ethicizing tell us abaout the status of morality and of social theory generally? Many, particularly philosophers, would argue that it tells us very little (Singer, 1981; Trigg, 1982). They would argue that to assume otherwise commits a gross violation of Hume's law. One would be going from matters of fact about the evolution of morality to matters of value about the actual truth status of morality.

Grant, it will be allowed, that epigenetic rules did come into place and action as just supposed. This tells nothing about whether or not the morality that we sense through the epigenetic rules has any independent validity. Consider analogously the state of the physical world. Eyes, hands, noses and so forth certainly came through evolution, but is this to deny the ultimate validity of an independently existing real world? Obviously not, would be the critics response. Consequently, analogously, nothing yet has been said about the status of social claims. To suppose otherwise is to plunge us right back into Victorian progressionism.

However, the sociobiologists argue, and today, after some hesitation, I totally agree, that thus to dismiss the relevance of the evolution of morality for an understanding of the status of morality is altogether too quick and negative. What is argued by sociobiologists is that once one has grasped that morality is the product of a long, directionless process which has its being solely in its adaptive nature for humankind (taken individually), then we must accept that morality is simply a subjective

phenomenon, with no being or reality outside the human dimension. That is
to say, given the Darwinian evolution of morality, we can positively
exclude alternative positions such as that morality resides ultimately in
God's Will, as is argued by Christians; or in transcendental archetypal
forms, as is argued by the Platonist; or in non-natural properties, as was
argued at the beginning of this century by G. E. Moore (1903); or even as
the conditions of rational beings interacting, as was argued by Immanuel
Kant (1959). Morality is, as the philosopher David Hume (1978) himself put
it, a matter of sentiment. We think morally because it is in our
biological interests to do so, but it has no being or meaning beyond this
(Ruse, 1984b).

How then does one account for the fact that so many people are
mistaken about the true nature of morality? Why is it that so many, even
when the evolutionary backing is explained, nevertheless refuse to accept
the consequences? Sociobiologists have already answered to this important
question. They agree that, phenomenologically speaking, morality has an
objective dimension. It is undoubtedly the case that when one makes a
moral statement, for instance, "Do not kill", the presumed backing is not
simply the subjective whim of the speaker. If I tell you not to kill, I am
not simply saying "Please don't kill because I don't like it", or "Killing
upsets me", or even that "Killing makes me unhappy". Rather, when I speak
morally in such a way, what I am doing, or at least what I think I am
doing, is appealing to objective eternal truths about the ultimate status
of morality. I am saying "Do not kill, because killing is objectively
wrong".

Morality is not just simply a question of my wishes, or your wishes,
or anybody else's wishes. Even if none of us thought about it, killing
would still be wrong, just as much as the Eiffel Tower would continue to
stand, even if all Frenchmen were asleep at some point and no one were
looking at it. In other words, the Darwinian argues that although morality
is subjective, we humans "objectify" it, meaning here that we give it an
apparent objective dimension (Mackie, 1977).

Why should this be so? Simply, argues the sociobiologist, because
unless morality had such a pressing objective-like dimension, it would not
command our attention. Suppose that I be faced with a decision, and from a
biological point of view it is in my interests to help you. However, this

biological interest is concealed from me, and what I am aware of are my surface feelings. Under normal circumstances, no doubt, I have absolutely no inclination, or desire, or intention, or anything whatsoever of helping you. Why, for instance, should I jump into a rushing river, thereby putting my own life at stake, simply to help you? However, what urges me on is my moral sense - that I have a moral obligation to help you. But note, this moral obligation must be something more than simply a feeling of wishing or otherwise, because the simple fact is that I do not wish to help you. This sense of moral obligation must be something which drives me on, **despite my normal feelings.**

In other words, I save you because it is right to do so, not because I want to save you, or because you want me to save you. I save you simply because that is the moral thing to do. In other words, morality has its effect upon us, its prescriptive effect upon us, because it has this apparently objective dimension. But, argues the sociobiologist, this objective dimension is itself part of the adaptive ploy that constitutes morality, and although we objectify our moral claims, in reality they have no such ultimate objective reference.

Do note that Hume's law is not being violated. It is not being argued that moral claims can be deduced or identified with factual claims. Rather, what is happening is that moral claims are in some sense being explained, or perhaps even more precisely explained away, by factual claims. Do note also, however, the sense in which "explained away" is being used. No one is denying morality. What is being denied is that morality has some ultimate objective reference, such as the Will of God, or whatever.

You may perhaps still find this conclusion somewhat difficult to understand and accept, so let me conclude this section by pointing out that if you take seriously the notion that evolution is not directed, then you must also take seriously the possibility that our morality might have been quite other than it is. For instance, had we not involved from savannah-dwelling primates, but from cave-dwellers or something like that, then our sense of morality could well have been very different from that which it is today (Ruse and Wilson, 1986). For instance, we might have been analogous to the termites, and thought it a highest moral duty to eat each other's feces. (Termites molt periodically, thereby losing certain essential parasites. In order to reinfect themselves, they eat each other droppings).

Indeed, as pointed out earlier, we might have had no moral sense whatsoever, and simply have been disinterested, calculating beings. But, this is so, we must therefore open up the genuine possibility that we could have evolved in such a way that, even if one supposed an objective, separate morality, we would have been totally unaware of it. This, it seems to me, is a contradiction of objective morality as it is usually conceived, for if God's Will, or whatever, is supposed to be anything, it is supposed to be binding on us humans. In fact, the situation is even worse. It might well be, given objective morality, that God wishes us all to hate each other, but it so happens that randomly we have evolved the other way, and thus believe that morality is liking each other. Of course, these are ridiculous possibilities, and something must go. If one is not to reject Darwinian evolutinary theory, then I am afraid objective eternal morality must be discarded.

VII. RELATIVISM ?

We have before us now the main outlines of the modern evolutionary approach towards a deeper unterstanding of social theory. The old Spencerian evolutionary ethics is a thing of the past, but the new Darwinian-inspired sociobiological understanding of ethics offers, I believe, an exciting new dimension to our understanding of ourselves. Let me conclude this discussion by raising and looking briefly at three points which may well have occurred to the reader. I shall thereby be able more fully to explain the sociobiology of ethics, and perhaps put to rest some natural queries and worries.

First, it may be feared that since everything is being put back on human nature, and since it is being explicitly affirmed that ethics is not objective in the usual sense of the term, that what we have is simply a brand of relativism. We no longer have cultural relativism, the old idea that whatever a society deems acceptable is thereby morally legitimated, but we seem to have a newer version, with a biological veneer. Apparently morality is now simply relative and so long as your genes tell you that it is acceptable to do something, then it is really acceptable to do so. If,

for instance, I have an inclincation to rape small children, then because
this is my natural inclination, it is perfectly acceptable. Morality
collapses into mere likes and dislikes, and as such is no true morality at
all.

One's natural worry about this conclusion is quickly answered by
pointing to our shared human nature. Human beings are indeed different:
some are tall, some short, some are black, some are white, some are male,
some are female. Nevertheless, we are united by far more than we are
separated. We are members of the same interbreeding gene pool and as such
share a common heritage and, to greater than lesser extent, a common
nature. This being so, although indeed morality is relative to human
beings (and perhaps some organisms close to us) this is in no way to say
that morality is in itself relative, meaning that what one person believes
another might not, and so on. We still have clear standards within the
human species, and this is all that one really needs for morality. Jesus
told us to love our neighbours as ourselves. He did not tell us to go out
and love the planet Mars.

Moreover, note that morality only works if it is something which is
shared. If I have a moral sense, but you do not, then perhaps I end up
being a saint; but, I also end up leaving no genes behind me (Mackie,
1978). Morality works because we are all in the same boat together, as it
were. Consequently, the sociobiological analysis of human moral thinking
and action is one which presupposes, and reaffirms, the internal stability
of the moral sense. It is not something which would have everything
collapse into some form of relativism. Of course, none of this is to deny
that perhaps certain broad behavioural differences might emerge between
races, and this might even lead to certain moral differences between
members of different races. However, one feels sure that any differences
will be no more, and probably far less, than can be brought about fairly
readily by cultural manipulation or other forces. In any case, what we do
know biologically is that peoples from different lands are far more
similar than they are different, and there is no reason to think that this
will be otherwise when it comes to our moral awareness.

VIII. RELATIVES, FRIENDS, AND STRANGERS

Second, let us dig a little more deeply into the question of whether or
not the evolutionary approach really does yield a social perspective
identical to that which is endorsed by modern philosophical and political
theorists. Is it really the case that our animal nature and our
philosophical nature thus neatly coincide? A query that one might well
have centres on the mechanism of reciprocal altruism. It will be
remembered that this only works if one has expectations of returns, of
some fashion. Might it not be argued that when there are no returns, one
expects to find the moral sense turned off, and is this not in some way a
negation of the very nature of the moral sense? Morality demands that we
give without counting the cost, not that we give expecting some response.

However, this query is fairly readily answered by the evolutionist.
Certainly, we expect some form of reciprocation, and if no reciprocation
is forthcoming, we expect to find that people's feeling of moral
obligation will in some way diminish, or at least change. However, this is
not to conflict with true moral thinking. In fact, what we find is that
moralists, no less than anyone else, feel little obligation to make mere
suckers from themselves - that is, to being used without hope or
expectation of return. Suppose I do something for you, and it might
reasonably be the case that you could do something for me, yet you refuse
to do this. Suppose, for instance, I have helped you one year when you
were starving, and now this next year my crop has failed, whereas yours is
abundant. Suppose also that you refuse to help me. Then I can certainly
turn to you and ask you to help me, not because I want you to help me, but
because it would be moral for you to help me. In other words, thanks to
morality we have the lever over others that we can expect and enforce
returns, in the name of morality.

However, having said this, there is at least one point where I see at
least some tension between a biologically backed moral understanding and
what moralists and social theorists frequently argue. This arises over the
range to which our moral sense and obligations extend. It is traditionally
argued by moralists - and frequently claimed by theorists - that we have
an equal moral obligation to each and every individual (Singer, 1972).
Kant, for instance, exhorted us in his categorical imperative to treat

humans as ends in themselves, but he did not, for instance, say treat those humans that you know as ends in themselves, but all strangers as means, or anything like that. Likewise, the position of the utilitarians is that one ought to maximize happiness and minimize unhappiness, of all. John Stuart Mill (1910), the most famous of all the utilitarians, explicitly stated that each individual counts for one, and that no individual counts for more or for less.

My feeling is that the Darwinian certainly can accept much that is said here; nevertheless, there is going to be a certain gradation, not merely in feeling, but in awareness of moral obligation. In particular, the Darwinian will argue that we have the strongest sentiments and obligations to our closest kin, then to those who are at least within the same pool in some sense as we, and only after this to others. What I mean in particular is that because of kin selection, the Darwinian will probably feel that his or her obligations are greatest to his or her children. Thus, for instance, when faced with the possibility of spending fifty dollars on food for one's own children, or sending fifty dollars to Oxfam, even though perhaps the money sent to Oxfam might feed say five people, and the money spent on one's own child will feed only one, one would feel nevertheless the obligation to feed one's own child. Likewise, because reciprocal altruism only really works in an situation where there is at least some potentiality for return, the Darwinian would, I suspect, feel more of an obligation to support say the helpless and poor in one's own society than those in say distant lands, like Africa.

Do not misunderstand me. I am not saying that the Darwinian would help only those who could possibly help in return. For instance, I am not saying that the Darwinian would feel no sense of moral obligation, say, to paraplegics in one's own society. This conclusion shows that you have missed the whole point. Morality takes one above the immediate wishes or desires, so that one will aid the helpless. Certainly, behind the veil of ignorance as it were, one can imagine being in need no less than are others. Nor am I saying that present distribution of funds (say, between the needly of our own land and those of foreign lands) is acceptable. And, I am certainly not saying that one should just simply spend all of one's money on one's own children, and nothing on anybody else. Nevertheless, I cannot help feeling that there is going to be a certain sense of differential. This, of course, patterns what we actually do think and do. We do, in fact, spend far more time and effort on our own immediate family than on the families of others. Often it is said that this is simply wrong. The Darwinian would quite flatly contradict this.

Why does Darwinism differ from what many social theorists would argue at this point? I suspect that much of the difference comes about because technology has outstripped our human abilities and propensities. Until recently, certainly until a century or so ago, it simply did not make much sense to talk about moral obligations to strangers in far lands. If one was an Englishman, say, living in England at the beginning of the nineteenth century, one could hardly talk about moral obligations to some undiscovered tribe in darkest Africa. Only inasmuch as one started to impinge on strangers did any kind of moral question arise. There was certainly no question, say, of air-lifting food to starving Ethiopians a hundred and fifty years ago. Consequently, moral obligations really did begin at home. Jesus tells us love our neighbours as ourself. The problem is one of deciding who, precisely, is our neighbour. Until recently, our neighbours really were those people with whom we came into fairly direct contact, and this was certainly even more the case at the time when our moral awareness evolved. Now, thanks to technology, the whole world ist part of our moral pool in some ultimate sense, but our moral abilities have not really extended that far.

IX. PROSPECTS

What this all means is that we have got to sit down and rethink important questions about problems which are facing today's world and how to approach them, given our human nature, which evolved in rather different circumstances. Which task brings me to the third and final point I want to raise, namely that to do with the way or ways in which a knowledge of the biological underpinnings of our moral and social nature leads to possibilities of change or other fresh approaches. You might still think that despite all that I have argued to the contrary, the sociobiological approach to moral and social theorizing is simply social Spencerianism by another name. Perhaps, indeed, this approach has its roots far farther back in time, even going back to the arguments of Thrasymachus in Plato's

Republic. You might think that, now we have seen that the evolutionist admits our obligations to strangers might not be as great as those within our group in some sense, it can be seen that Darwinism is simply arguing for a legitimatization of the **status quo.** Basically, the position of the evolutionist is that those who have material goods should hang onto them, and those who do not are unfortunate but must do without. Everything in the present state of affairs is given a validity and backing by the evolutionary process.

In other words, the sociobiologist is simply arguing that that which has evolved is that which is natural, is that which is good, or more crudely, one is arguing for some form of "Might is right". Indeed, with Thrasymachus one is suggesting that once we have insight into human nature, we can manipulate it to our own ends. We see that humans are deluded by their genes, into accepting some sort of morality. However, once we realize this, that morality is simply a delusion imposed perhaps not by a ruler or by a social contract sometime in the past, but by the evolutionary process, there is absolutely nothing to stop us from breaking free from this delusion and operating life to suit our own ends. In short, now that we see that morality is simply subjective, there is nothing to stop us from turning right around and escaping from morality and behaving as we will, namely, behaving in a totally selfish way. Why should I not, for instance, go out and seduce young girls?

Normally you would say that I should not do this because it is wrong because it violates the Categorical Imperative or some such thing. Now I realize that belief in the Categorical Imperative is just an adaptation, like hands and eyes, and that if in fact I want to go out and seduce young girls because I enjoy doing it, or some such reason, then there is absolutely no reason why I shouldn't do this.

However, I would argue that the evolutionist is not committed to this horrendous conclusion at all, nor to anything like it. First of all, and most importantly, we are ourselves humans, and we have the biological nature that others have. Consequently, we cannot simply throw off our morality just like that, as we might perhaps do if morality were simply a social convention imposed from on high by someone else. We are moral beings, with our own sensitivities and wishes and likes and dislikes. Hence, being aware of the nature of morality does not mean that in any sense we can stop being moral. If I went out and started seducing small girls, I would feel extremely tense and unhappy and conscience-stricken. Even thought I might know that this is, itself, purely, a biological reaction, such knowledge would not necessarily eliminate it. The fact that pain is caused by cancer does not makes pain itself any less unpleasant.

Of course, you might argue to the possibility of some sort of wholesale genetic engineering to eliminate morality, or some such thing, but I really cannot see why one should even want to attempt this. Are we to stop living socially, or is the suggestion that we might live socially in some much better way? If so, then an argument must be mounted. But not otherwise. In any case, I suspect that attempts at wholesale genetic engineering in order to alter our moral sensitivities might prove much more complex and difficult than most optimists suppose. Apart from anything else, there is deep realization by biologists that genes do not exist in isolation from each other. Hence, if you alter one physical or psychological characteristic, you are liable to alter a great many more. We might, for instance, eliminate our moral sense, but find at the same time that all other sorts of desirable things have gone, like an ability to enjoy oneself.

Nevertheless, having said this, this is not to say that, in the light of our new knowledge about our biological nature, we are totally helpless, or that there is nothing that we can aim to do to work within the system and improve the lot of ourselves and of our fellow humans. Take for instance the question of overpopulation. Most people in the west find thoroughly repugnant the Chinese attitude towards population control where people are virtually forcibly sterilized, abortions are imposed upon women very late in pregnancy, families are rigorously confined to one child, and so forth. To us, behaviour like that is a gross infringement upon personal liberty. Nevertheless, at the same time we are aware of the dreadful problems which are being caused by overpopulation in countries like China. Any material advance is immediately wiped out by yet more mouths to feed.

What I would suggest is that our feelings of repugnance come about because, under normal circumstances, namely in a group where population is not a great problem, as was the case in our past, there were strong selective pressures towards the cherishing of individual liberty. Now, we find ourselves, because of improved medicine and so forth, in situations where liberty has simply got to be constrained for the value of the group good. In other words, we can see the need to suppress some of our more immediate moral yearnings, in order to satisfy a more reflective and general recognition of what is right and wrong. As I have said, I would not argue that simply becoming aware of what is going on makes all well. Nevertheless, it is surely a first step towards an understanding of the problems that we face, and perhaps even a resolution. Perhaps through education we can come to a more sympathetic realization of the new difficulties that we face in life, thanks primarily (as noted above) to technology.

If I were to say any more, I would have to say a great deal more, and that would make this discussion overly long. Hence, I will rest my case here, simply hoping to have made you aware of the fact that an evolutionary approach to social problems does not thrust us back in the nineteenth century. It can, in fact, open the possibility of a fresh approach to life's problems.

X. CONCLUSION

Biological theory and social theory have had a troubled relationship in the past, but we must not remain prisoners of this past. Rather, we must use those powers that evolution has bequeathed to us peer back into our own nature and to look forward to a happier future. I argue that one way it is possible to do this is by bringing to the fore the biological substratum of our moral and social propensities, and by then acting upon this new awareness. Given the problems which face humankind today, there is no time to lose. Let us therefore start to act at once.

FRANZ M. WUKETITS

EVOLUTION, CAUSALITY, AND HUMAN FREEDOM.
THE OPEN SOCIETY FROM A BIOLOGICAL POINT OF VIEW*

I. INTRODUCTION

This paper is a critical examination from a biological viewpoint of the
notion of the open society. It aims at giving a brief account of some of
the most important ideas of contemporary biology - ideas which could
possibly consolidate at least the basic postulates of an open society, a
free society. I shall concentrate on the concept of causality in modern
evolutionary theory and shall try to outline a systems view of evolution.
According to such a view the evolution of living systems, our human one
included, is not determined by "pre-existing" laws (for example Leibniz'
pre-established harmony) nor is it, in the long run, just a series of
random events. Evolution is rather a self-organizing, a "self-planning"
process.

This approach to understanding the evolving biosphere goes beyond
determinism and indeterminism - and it may rescue something of the old and
venerable idea of human freedom. In an "open universe", only man, the
rational animal, can plan his future by learning from his own history. If
he takes the opportunity to use "self-planning", he can experience
freedom; if not, he could develop towards his self-destruction.

Warning: Biological concepts and theories often have paved the way to
dangerous ideologies. Biology has sometimes been confused with
"biologism", **e.g.** Social Darwinism. Therefore, by trying to describe human
systems in biological terms, we must remove biology from any ideological
claims. Ideological doctrines such as Social Darwinism stem from
misreadings of certain biological theories. Such misconceptions in
themselves spell the end of an open society. Karl Popper's **The Open
Society and its Enemies** comes to the following conclusions:

49

M. Schmid and F. M. Wuketits (eds.), Evolutionary Theory in Social Science, 49–77.
© 1987 by D. Reidel Publishing Company.

> ...to progress is to move towards some kind of end, towards an end
> which exists for us as human beings. 'History' cannot do that;
> only we, the human individuals, can do it; we can do it by
> defending and strengthening those democratic institutions upon
> which freedom, and with it progress, depends. And we shall do it
> much better as we become more fully aware of the fact that
> progress rests with us, with our watchfulness, with our efforts,
> with the clarity of our conception of our ends, and with the
> realism of their choice. (Popper, 1962, vol. 2, pp. 279-80).

Likewise Julian Huxley, a grandson of Thomas H. Huxley (Darwin's
"bulldog") and himself an eminent evolutionist, in his essay 'Evolution,
Cultural and Biological' pointed to the significance of studying human
history "by clarifying the role man's ideas of destiny have played during
past cultural evolution, (in order to) make it easier for man to achieve
his true destiny in the future" (1957, p. 92).

One might ask what Popper's and Huxley's works have in common.
Popper's approach to understanding human affairs grew in the first place,
from his study of history and social philosophy, whereas Huxley, was
principally trained as a biologist.[1] I have put these quotations together,
however, because the message of both is rather the same: **It depends on us
to plan our own future.** And what I want to express in the present paper is
that the only possibility for us to achieve our true destiny is **to learn
from our evolutionary past**, that is to say, to learn from evolution as a
natural process.

I shall try to show what we in fact can and what we **should** learn from
evolution in order to gain perspectives for our future development. If we
have the right notion of evolutionary processes then we should be able to
ask the right questions of ourselves and of our society. I am going,
therefore, to explain what the "right" notion of evolutionary change in
the living world - and, thus, in the development of man as a biological
species - might be. But I am **not** going to present a sociological theory.
My concern here is to give a sketchy treatment of some ideas of
contemporary biology, ideas which could be of some interest to
sociologists.

In discussing the biological basis of social action we must avoid
extrapolations such as Social Darwinism. On the other hand, biology does
help us to better understand the origin and evolution of human social and
cultural systems. Biology, and particularly evolutionary theory, offers an
understanding of the **preconditions** of human social and cultural behaviour.

Hence **biosociology** means the attempt to conceive the biological
foundations of sociality (see Meyer, this volume) and not on admission to
reduce human social (and cultural) systems simply to biological
categories. Biological determinism, undoubtledly, is one of those ideas
that have harmed mankind.

II. THE SYSTEMS-THEORETIC APPROACH TO EVOLUTION: DARWIN AND BEYOND

It is fairly well known that Charles Darwin explained the phenomenon of
evolution by natural selection and that he like some other 19th-century
evolutionists such as Herbert Spencer took an "adaptationist" view.[2]
Although Darwin himself became aware of the incompleteness of the theory
of natural selection, this theory has had great influence not only on the
biological sciences but on human and social sciences as well.

Darwin's epigones, particularly the German zoologist Ernst Haeckel,
welcomed the explanation of living beings and of the diversity of life **via**
natural selection rather enthusiastically. And up to now the notion of
natural selection has been the basic idea in evolutionary biology and has
become almost a commonplace in discussions among biologists. During the
first four decades of the 20th century the concept of selection - and, in
general, Darwin's view of evolution - was much supported by genetic
studies. These studies, together with those in paleontology, biogeography
and other fields, finally led to what Julian Huxley called the "modern
synthesis" of evolutionary biology.

To be sure, there are differences between "classical Darwinism" and
today's synthetic theory, due to advances in genetics, molecular biology,
biochemistry, and so on. Nevertheless, selection from the viewpoint of
synthetic theory still works as an **external evolutionary** factor and
synthetic theory expresses a firm belief in a mechanistic world order. In
opposition to such a belief, evolution is sometimes considered to rely
upon **internal**, "intraorganismic" principles. Riedl (1975, p. 89) lists 26
biologists and philosophers (from 1876 to 1969), who stress the meaning of
internal factors in evolution. They are all convinced that the
peculiarities of living systems cannot be sufficiently explained in terms
of Darwinian selection. In some cases, however, the internal factors of

evolutionary change are conceived similarly to the vitalist's conceptions of life forces (remember, for instance, Henri Bergson's **elan vital**).

The controversy between "internal" **versus** "external" principles therefore divides biologists into a party of mechanists and a party of vitalists. The acceptance of internal factors **prima facie** seems indeed to be at variance with a theory supposing natural selection as the only "driving force" of evolution.

A way out of this dilemma was the discovery of feedback-principles in living organisms and the application of cybernetics and systems theory to biology.[3] Systems thinking in particular has helped us to adopt a view going beyond "internalism" and "externalism". Recently Riedl (1977, p.358) has emphasized that "there is no inner or outer evolutionary mechanism that is allowed to operate independently." Thus, we have to take into account **systems conditions** "which link different levels of complexity to feedback loops of cause and effect" and which "are responsible for the evolution of life" 1977, p. 358; full details in Riedl, 1975; see also Ott **et al.**, 1985; Wagner, 1983; Wuketits, 1981; 1982b).

The argument runs: Selection neither operates merely as an outer principle (environmental selection), nor does it work solely as an intraorganismic factor. Evolution is directed, so to speak, by both external and internal selective agencies. We admit that selection is the major agent of evolutionary change, but we propose that it does not work in such a way as has been claimed by the proponents of Darwinism, neo-Darwinism and synthetic theory. In addition to Darwinian selection we maintain that the organism itself promotes a kind of selection. That is to say, "selective forces" are initiated by functional constraints of the organism (Gutmann and Bonik, 1981) and that there is a constant flux of cause and effect from one level of the organismic organization to another, **e. g.** from the cellular to the molecular level and **vice versa**. (On a multilevel approach to evolution and selection see also Gould, 1982.[4])

Is, then, Darwinism wrong? The answer is **no**. An improved version of natural selection, as proposed by the means of a theory of systems conditions, however, results in an extension of the Darwinian framework: The theory of natural selection needs to fit into a new conceptual scheme and to be incorporated into a systems-theoretic view. This view boils down to an assertion that highly organized systems such as biological organisms can only be explained with reference to **feedback causality**.

The conception of feedback causality or, as it were, "network causality" (cf. Riedl, 1975; Wuketits, 1981) replaces the one-way-causal paradigm inherent in mechanistic explanations of evolution since Darwin. Riedl (1977, p. 366) points to the consequences of the concept of feedback causality in biology:

> If it is true that feedback cycles can connect levels of different complexity ..., then we must accept a flow of cause and effect in two directions, up and down the pyramid of complexity. Then we should also accept causality in living beings as a system in which effects may influence their own causes. Biology would then at last accept causality as a network rather than as one-way chain, a notion which has been fundamental in physics since Galilei and Newton and a truism in social sciences, but which is still badly hampered in biology by its hundred-year-old entanglement in doctrines ...

Feedback causality, then, is an important issue in biological science. It throws new light on the problem of **teleology.**

The problem of plan and purpose in living systems has given rise to many controversies in biology. One might suspect that teleology is contradicted by causality and that teleological explanations are at variance with causal accounts. Since Darwin, teleology has been denied in circles that claim to solve the problems of life by means of the theory of natural selection. On the other side, all those who have demanded internal principles in evolution have tried to rescue teleology as a fundamental character of organisms. Thus, for some biologists, the notion of teleology is an emblem of outmoded styles of thinking, for others it is a principle indispensable to biological explanations.

One cannot ignore the fact that organisms, in one way or other, exhibit goal-directed activities and that there is a tendency in living systems to self-regulate, to maintain certain charateristics in spite of environmental fluctuations. And in embryology and development, goal-directed phenomena are striking. Most recently Rosenberg (1985, p. 37) has made the point:

> ... the more we know about the details of development in simple systems and complex ones, the more striking the phenomenon seems. The reason is that not only do whole organs and limbs develop in accordance with a plan, with a goal that is reached even in the face of interference and obstacles, but the component tissues and cells also differentiate and develop, and sometimes even regenerate, in an apparently goal-directed way.

To be sure, there is nothing mysterious about teleological phenomena
in organic nature, if - and only if - we take organisms as **complex systems
of interactions** between their elements and subsystems. Teleological laws,
then, can be conceived of as sets of causal laws or as principles of
feedback causality. Rosenberg (1985, p. 53) writes:

> Most attempts to show that such laws exist and that they demystify
> teleology turn on notions introduced in the development of
> cybernetics and general systems theory: **feedback loops** and **feed-
> forward loops.** The idea is that systems that appear to be moving
> toward a goal are controlled by ordinary causal interactions that
> are often large in number ... and that interact to produce the
> **appearance** of purpose in the eye of the observer. Such systems
> have been described as "directively organized" ones. Well-known
> examples of such directively organized systems and their goals
> studied in physiology include the body and its goals of
> maintaining an internal temperature of 37° C and maintaining water
> balance.

It seems to me, however, that Rosenberg disregards the notion of feedback
causality to some extent. Ordinary causality, the notion of causality
predominant in classical physics and mechanics, does not suffice to
explain the existence of feedback loops and feed-forward loops. Feedback
causality cannot be simply reduced or "cut" to a number of causal chains,
for it is a property of systems with high degree of organization.

A system trying to stabilize and to maintain its own properties
depends on the **functionability** of its elements and/or subsystems which,
conversely, rely upon the stability of the whole system. If you "cut up"
the whole system, the system's specific properties will disappear.

As a matter of fact the function (purpose) of any single element of a
system subsystem is determined by the system itself. At the same time the
system is not viable if it does not interrelate with the functions on the
level of its subsystems. These interactions between the whole and its
parts constitute feedback mechanisms and represent, some way or other,
internal selective principles. On condition that specific functions are
performed at all levels of organization, the whole organismic system
shows teleological behaviour.

Teleology, however, from the viewpoint of cybernetics and systems
theory no longer remains a matter of mysticism, for it is not to be
confused with **goal-intended** activities demanding a "designer" or anything
like that. Biological systems are only goal-directed, that is to say that
there is a certain direction of processes **a posteriori**, but no "ghost"

intending a certain end-state **a priori.** The goal-directedness in living
systems - particularly in embryology and development of the individual
organism - depends on a genetic program which is itself the result of
evolution. Any teleological process depends on two components: First, the
necessity of a specific function in order to make possible the survival of
the organism; second, the regulation of the function based on a specific
genetic program.

Teleology, therefore, is not a supernatural principle, but one,
rather, to be handled in terms of evolutionary biology. In order to avoid
confusion with metaphysical claims present day biologists and philosophers
of biology prefer to speak of **teleonomy** instead of teleology (cf. Lorenz,
1973; Mayr, 1974; Mohr, 1977; Wuketits, 1980, 1981, and others).[5]
Organisms, by definition, are **teleonomic systems** operating on the basis of
an optimized (genetic) program. "Teleology", then, means simply the
existence of "structures, processes, or patterns of behaviour which are
valuable or necessary for the self-maintenance of an organic system, that
is, an individual system or a system above the individual level **e.g.** a
population" (Wuketits, 1980, p. 283). The next question is whether or not
evolution on the whole is a teleological process. Is there - or, could
there be - a principle directing evolution in the long run towards a
certain goal? And if so, what would that goal be? One point can be made at
the outset: There is no scientific reason to believe that the evolution of
living systems is rigidly directed towards any goal or end-state.
Evolution is an "open" process. Only the genetic program of an organism
(and of any species) contains goal-directedness towards survival; but
there is no mechanism known, or even imaginable that could anticipate any
"final result" of evolution. These assertions, I think, require some
further elaboration.

The cornerstone of Darwinism is environmental selection. Modern
synthetic theory amounts to an assertion that this kind of selection
together with random mutations is sufficient to generate evolutionary
"progress". External selection as we have seen is not in itself enough to
direct living systems towards survival. Nor is the theory sufficient to
explain the tremendous order in living systems and the diversity of life.
External selection works short-sighted, so to speak, and it is hardly
intelligible that evolution has come to pass merely by means of short-
sighted selection and blind chance (mutations). According to Bertalanffy:

today's synthetic theory of evolution considers evolution to be
the result of chance mutations, ... of 'typing errors' in the
reduplication of the genetic code, which are directed by
selection, **i.e.** the survival of those populations ... that produce
the highest number of offspring under existing external
conditions. (1973, p. 160).

Moreover:

In contrast to this it should be pointed out that selection,
competition and 'survival of the fittest' already **presuppose** the
existence of self-maintaining systems; they therefore cannot be
the **result** of selection. At present time we know no physical law
which would prescribe that, in a 'soup' of organic compounds,
open systems, self-maintaining in a state of highest
improbability, are formed. (**ibid.**).

Bertalanffy died in 1972 and therefore, of course, was not abreast of
recent developments in biophysics concerning the origin of living systems.
Bertalanffy was right to criticize synthetic theory but he was not right
as to his demands on physical laws. Certainly, there is no physical law
which (in any defensible sense), **prescribes** that living systems be formed.
But matter has a general tendency towards **self-organization.** Manfred
Eigen's model of the **hypercycle** deduces the emergence of self-maintaining
systems from material self-organization, and it has been shown that
selection already occurs at prebiotic levels of evolution. The model of
the hypercycle has been exhaustively described by Eigen and his
collaborators (see Eigen and Schuster, 1979). In short, a hypercycle is an
autocatalytic system representing a cyclic arrangement of nucleic acid and
proteins. Thus, the model of the hypercycle is a systems model of
prebiotic evolution containing a formula of preconditions to life.

Self-organization may be supposed as general principle in the
unfolding of various systems from those of elementary particles up to
social and cultural systems (cf. Wuketits, 1982a). This is not to say that
complex systems are to be reduced to more simple ones, but only that the
development of systems at various levels of complexity depends on similar
universal principles. According to game-theoretic models the principle of
self-organization allows us, moreover, to understand how and why
fundamental physical laws of nature are integrated into higher functional
levels (cf. Leinfellner, 1984). The major features of evolution, then, are
increasing complexity by integration and the **emergence of a great variety
of systems.** Evolution, particularly in the organic world, has to be

characterized by increasing complexity and diversity of systems. (Fig. 1 is a rather quick draft of these properties of organic evolution).

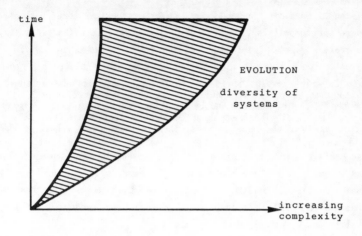

Fig. 1

Increasing complexity of systems on a hierarchically organized world is an expression of self-organization. To the extend that self-organization obviously is a general systems property of matter, we are able to particularize physical principles that allow for the emergence of highly organized self-maintaining systems like organisms.[6] Hypercycles as explained by Eigen and Schuster (1979) represent a new level of organization in addition to entailing the prerequisites of natural selection:

> Noncoupled self-replicative units guarantee the conservation of a limited amount of information which can be passed on from generation to generation. This proves to be one of the necessary prerequisites of Darwinian behaviour, i.e. of selection and evolution. In a similar way, catalytic hypercycles are also selective, but in addition, they have integrating properties, which allow for cooperation among otherwise competitive units. (Eigen and Schuster, 1979, p. 6)

Hence it has been shown that under special conditions evolution at the molecular level leads to the emergence of organic systems or, to put it more precisely, that those physical properties of matter that under suitable environmental conditions allow for metabolism and self-reproduction prove to be sufficient for evolution by natural selection.

However, the model of hypercyclic self-organization amounts to a revision of Darwin's theory as well as of synthetic theory. Both theories account for **competition**, but evolution, moreover, is to be explained as a "cooperative game". Leinfellner (1984, p. 256) comes to the following conclusion concerning the incompleteness of Darwin's theory:

> If Darwinian evolution is the continuous genetic adaptation of organisms (systems) or species to the changing environment by the forces of selection ... and by mutation, where mutation is simply the main variation-creating force and selection is the filter screening off the less fit ones, then it tells us only half of the story.

Consequently, Leinfellner (1984, p. 256) has developed a **dynamic** theory of games which "unifies all these factors or causes in a general system of **mutual** causation, including **competitive** and **cooperative** games" (my emphasis). All this fits into the above-mentioned conception of feedback causality and a systems theory of evolution.

Since hypercycles represent systems of interaction between at least two elements - DNA and proteins - , it is apparent that even at the bio-molecular level of organization the one-way-causal paradigm is insufficient to explain emergence and complexity.

Let me now summarize the systems-theoretic approach to evolution and compare it to Darwin's theory and synthetic theory.

(i) Classical Darwinism as well as modern synthetic theory insists on external selection and on competition, whereas from the viewpoint of systems theory the evolutionary process depends as well on internal (intraorganismic) selective forces and can be described as a dynamic game unifying competitive and cooperative components.

(ii) Any organism is a highly organized self-maintaining open system. I suspect that the Darwinists have not done justice to this aspect. For to explain evolution with resort to environmental selection already presupposes the existence of living systems, that is systems able to evolve. But what about the origin of such systems? Only recently there have been attempts to reconstruct the origin of bio-systems according to a theory of general self-organization, taking cognizance of cyclic processes which, under very special conditions on primordial earth, led to a specific arrangement of biological macromolecules. This theory of self-

organization makes the evolution of life understandable as a systems
process.

(iii) Living systems are organized teleologically, that is to say they
perform functions of certain value for the sake of survival. The systems
view allows us to understand teleology (teleonomy) as a natural principle.
Even the first hypercyclic systems - organizing themselves about three
billion years ago - were in a way teleonomic systems, but in order to
explain this development we have no reason to appeal to any kind of
mysticism. In order to conceptualize teleology without metaphysical
notions it is important only to advance the concept of feedback causality.
Feedback or network causality is an epistemological prerequisite of an up-
to-date approach to evolution.

What do these biological conceptions mean for man and human society?
First of all, a dynamical systems view of organic evolution brings us
closer to the idea of an open society, to use Popper's term again, in so
far as from a systems view of evolution, there has been no "designer", not
even a physical law to prescribe that human evolution procedes towards any
goal or end-state. The human race has to find its own goal. Therefore, as
Leinfellner puts it,

> we have to give up the deterministic view in favour of a fuzzy and
> indeterministic leeway, which has the great advantage for human
> societies of guaranteeing a certain freedom of choice for the
> individuals. (1984, p. 233)

Indeed, systems view of evolution goes beyond determinism. This is its
significance to the study of human systems.

III. THE EVOLUTION OF MAN: BEYOND DETERMINATION AND DESTINY

Myths of destiny, for example the belief in eternal evolutionary laws,
have harmed mankind whenever they have been realized in political systems.
Any kind of determinism, any belief in destiny, is therefore, a dangerous
ideology (see **e.g.** Wuketits, 1985), acting against open society and human
freedom.

More than 2,000 years ago Plato agitated **against free society**:

> The greatest principle of all is that nobody, whether male or
> female, should be without a leader. Nor should the mind of anybody
> be habituated to letting him do anything at all on his own
> initiative; neither out of zeal, nor even playfully. But in war

and in the midst of peace - to his leader he shall direct his eye
and follow him faithfully. And even in the smallest matter he
should stand under leadership. For example, he should get up, or
move, or wash, or take his meals ... only if he has been told to
do so. In a word, he should teach his soul, by long habit, never
to dream of acting independently, and to become utterly incapable
of it. (Quoted after Popper, 1962, vol. 1, p. 7)

Some of us experienced the spell of Plato: during World War II under the
criminal reign of the Nazis.

On the level of biological systems, insights into feedback mechanisms
and interactions between different levels of complexity prompts us to re-
formulate and to re-evaluate the problem of determinism. These insights
unify random mutations, environmental selection and intraorganismic
selective forces (self-regulation) into a set of systems conditions so
that we need no longer ask "determinism or indeterminism?". "Chance or
law?" Evolution is neither predetermined by rigid laws nor based on pure
randomness, but is rather a self-planning process including both law and
chance (cf. Riedl, 1975, 1977; Wuketits, 1979, 1981, 1982b). This is quite
in accordance with Corning's theory of **functional synergism** (see Corning,
this volume).

Popper states that "universal laws make assertions concerning some
unvarying order " and says that evolution - be it in the biological or in
the sociocultural sphere is, on the contrary, "not a law, but only a
singular historical statement" (Popper, 1961, p. 108). Popper's intuition
is that evolution, although in accordance with causal laws (**e.g.** the laws
of mechanics), does not allow for one **universal** historical law. The modern
systems-theoretic approach to evolutionary phenomena offers convincing
evidence that Popper is right.

What does seem clear, however, is that rejecting determinism does not
mean asserting that evolution produces disordered systems. The products of
evolution are in fact highly organized systems and, thus, **systems of
order**. But this order is not the result of strict determinism, but of the
systems conditions of evolution. Thus, we have order without universal
laws!

This statement is supported by **nonequilibrium thermodynamics** and in
particular by Ilya Prigogine's concept of **order through fluctuation**. In
classical thermodynamics (equilibrium thermodynamics) fluctuations only
play a minor role. **Non-linear** systems, on the other hand, under conditions

far from equilibrium tend towards fluctuations which "can force the system
to leave a given macroscopic state" (Prigogine, 1976, p. 93). Consequently
the system assumes "a new state which has a different spatiotemporal
structure" (Prigogine, 1976, p. 93). Thus, the following relations obtain:

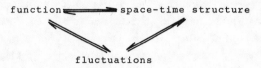

The system drives towards a new dynamic regime, so to speak, and is to be
characterized by a high degree of energy exchange with its environment. In
any case, nonequilibrium systems are open and self-organizing systems.

    Prigogine developed his concept of nonequilibrium thermo-dynamics
during the last three decades or so. (A comprehensive account of the
scientific and philosophical foundations and consequences of this concept
appears in Prigogine and Stengers, 1981.) The essay quoted above
(Prigogine, 1976) is a short treatment of self-organization and
nonequilibrium thermodynamics in human (social) systems. Erich Jantsch in
his **Design for Evolution** (1975) has also presented an extensive account of
self-organizing human systems. Jantsch is however unfortunately prone to
mysticism, and the whole enterprise seems, in last instance, to have to do
with "neo-romanticism" or something like that (see Wuketits, 1985).

    The concept of order through fluctuation is principially a physical
theory. I agree with Jantsch, Prigogine, and some others, that this theory
is applicable to all domains - from physics to sociology, but I do not
think it useful to try to describe social systems exclusively with resort
to order through fluctuation. Nor do I agree with Jantsch's assumption
that "all organization in the universe would be physical and psychic at
the same time" (Jantsch, 1975, p. 38). Apart from such speculations,
however, order through fluctuation is an important concept, one compatible
with the systems view of evolution and one that elucidates the preconditions of evolution in the human sphere.

    Finally, some words about the **irreversibility** of evolution. Systems
theory and the thermodynamics of open systems help us better to understand
why evolution, though not determined by rigid laws, can never retrace its
steps. It is one consequence of systems conditions of evolution that

certain structures and function are, as it were, cemented during the
evolutionary process. A dolphin, despite new environmental constraints and
new functional requirements defined by aquatic life, will never become a
fish. A bat will never become a bird, although external selection may
indeed nudge it in this direction (see Riedl, 1975; 1977). Again, there is
no principle that prescribes the direction of evolution. Since, however,
evolution is a **self-planning** process, certain lawful principles become
established "step by step". For man, the observer looking backwards,
evolution seems to be a process directed towards a certain goal. But when
we examine the systems conditions of evolutionary change, we soon
recognize that the pretended pre-established harmony is in fact a "post-
established harmony".

The evolution of man as a biological species is no exception to these
"rules" of evolution. Whenever we notice certain **trends** in evolution, be
they in the human or in the subhuman sphere, then these trends must not be
identified with pre-determined laws. Popper (1961, p. 128) clearly states
that trends - in biological and in socio-cultural evolution - do indeed
exist, but that "their persistence depends on the persistence of certain
specific initial conditions (which in turn may sometimes be trends)".
Popper continued by arguing against **historicism**, which is after all a kind
of determinism:

> This, we may say, is the central mistake of historicism. **Its 'laws
> of development' turn out to be absolute trends**; trends which, like
> laws, do not depend on initial conditions, and which carry us
> irresistibly in a certain direction into future. They are the
> basis of unconditional **prophecies**, as opposed to conditional
> scientific **predictions**. (Popper, 1961, p. 128)

All things considered, our view of the evolution of man goes beyond
determinism and destiny. Those who believe in predestination - be it in a
theistic sense or in the sense that historical laws are supposed to be
natural or social forces acting upon man - are in fact supporting
ideological claims that are at variance with the unfolding of a free
society.

IV. THE EVOLUTION OF MAN: BEYOND PHYSICALISM AND MENTALISM

There cannot be serious arguments against the theory that the human race
is a result of evolution. Strictly speaking this is not a theory but a

matter of fact: A wealth of fossil evidence indicates that human beings descend from other animals and that **Homo sapiens** - the only living species of the family **Hominidae** - developed from more primitive species.

There are, of course, still some unsolved problems in dating the origin of the hominids. However, assuming that the genus **Australopithecus** was a true hominid, it can be said that the first man appeared in the early Pliocene (about 4 million years ago) (see Washburn, 1980). Archaic **Homo sapiens** appeared some 100,000 years ago and modern **Homo sapiens** about 40,000 years ago. According to Pilbeam (1984) the species **Homo erectus** and **Homo sapiens** (including the Neanderthals who disappeared about 35,000 years ago) probably represent a continuum in human evolution.

My concern is, first, to say that the view that evolution proceeded in a straight, inevitable progression of species culminating in hominids and, particularly, in **Homo sapiens** is untenable. The emergence of man was a result of the systems conditions of evolution. The development of man became possible, but there was no necessity of progressing towards **Homo sapiens**. During the evolution of this biological species, however, man acquired his status as **animal rationale** - and this was a decisive step in evolution, because for the first time a living system could begin to monitor and control his own development. But how did man's peculiar **self-awareness** (including **death-awareness**) develop? Is the human **mind** simply a result of organic evolution? This is my second concern: to draw attention to an evolutionary approach to man's mental abilities.

Many people still find it difficult to explain human mental capacities (knowledge, reason, language, morality) in terms of evolution.

But among contemporary biologists there is consensus that the behaviour of organisms depends on anatomical structures and specific (physiological) functions. Thus since structures and functions are results of evolutionary processes so are behavioural patterns. Basically this applies to **human** behaviour. Even our mental capacities depend upon biological structures (brain, nervous system) and, consequently, have developed alongside these structures.

In the table below, the striking correlation of brain size with advances in cultural development (recorded by artefacts) leads to the assumption that man's mental abilities rely upon the complexity of the brain. (For some data on hominid evolution see table below. After Pilbeam, 1984, and Washburn, 1978)

Hominid	Approx. age, 10^6	Brain size, c.c.	Artefacts
Australopithecus afarensis	4	450	Primitive stone tools?
Australopithecus robustus, africanus	1.6	475	Primitive stone tools
Homo erectus	1 - 0.5	750-1300	Stone tools somewhat more complex; use of fire
Homo sapiens (Neanderthal man)	0.07-0.04	1400	New tool types and techniques artefacts in-dicate self-awareness
Homo sapiens (Modern man)	0.04	1500	Abundant stone and bone tools and other ar-tefacts; tech-niques beco-ming more com-plex

The evolution of the brain is a systems process; neurons have been linked together into complex patterns of organization. The organization of the human brain is the biological basis of mind. Thus, **mind is a systems property of the brain,** and biological evolution is the precondition of spiritual evolution. This is one of the basic theses of **evolutionary epistemology, i.e.** the evolutionary theory of (human) knowledge. (See Bartley, 1976, 1983; Campbell, 1974; Lorenz, 1973; Mohr, 1977; Plotkin, 1982; Popper, 1972; Riedl, 1980, 1984; Riedl and Wuketits, 1987; Vollmer, 1984; Wuketits, 1983, 1984a, 1984b)

From the viewpoint of evolutionary epistemology man's cognitive activities - and even such knowledge increments as learning and thought - are products of evolution. Moreover, evolutionary epistemology explains certain limits to our cognition and knowledge. We have to consider that our inborn cognitive apparatus developed under the circumstances of our evolutionary past. This apparatus was selected for the sake of survival in a world different from the world of today, which is mainly a product of man's activities. In the world of the australopithecines, **Homo erectus** and early **Homo sapiens,** the **ratiomorphic, i.e.** "pre-rational" algorithms that cannot fully cope with today's world were sufficient. The present day inadequacy of the ratiomorphic and yet existing cognitive mechanisms can be demonstrated by the conception of causality.

Human cognizance of causal relations has been programmed as the recognition of causal chains. Our innate expectation of linear causality, this is to say our inborn "cause-effect notion" (one-way-causal paradigm, see above), sufficed within the rather simple world of **Homo erectus** or that of any other of our phylogenetic ancestors. But it no longer suffices for complex phenomena; and it is inadequate for getting the present situation of man (which has become precarious, as we all know) under control. A look at any complex systems such as organisms, cultures, societies, or economic systems, makes it clear that such systems are built up of sophisticated patterns of interactions between many elements. To explain and understand the function of complex systems we have to adopt feedback causality. Since the human cognitive apparatus consist of more than innate "teaching mechanisms" (ratiomorphic mechanisms), we are able to transcend the archaic ways of our behaviour. For we are endowed with reason, we have the possibility of rational knowledge and thus, the possibility of understanding the existence of structures beyond our **mesocosm**.[7] This explains why we are able to overcome, at least to a certain extent, the biological limits to our cognition and knowledge. Much can be said for Levinson's evolutionary epistemology without limits:

> While our cognitive capacities indeed may have been selected based on their ability to perform well in our evolutionary environment, these circumstances of genesis need not restrict our ability to make sense out of environments with radically alien natures. (1982, p. 468).

I should say that the human cognitive apparatus is not yet free from evolutionary restrictions and that it never will be completely free from them because we are a biological species. However, reason has opened up completely new dimensions for us and we need not believe that our **rational capacities** will be limited for all time. I agree with Levinson that science - or, rational learning and thought, in general - has amplified and will amplify our world view. Levinson concludes that science and technology enlarge

> the evolutionary flexibility and range of our cognitive capacities by providing us with new sensory experience, speeding our powers of calculation and organization, and externalizing parts of our very cognitive processes themselves. (1982, p. 492).

Scientific knowledge has indeed led to an increase of our knowledge capacity beyond mesocosmic structures.

Bartley (1983) argues that we can properly understand the cognitive structures of other organisms and improve our own cognition:

> From the height of our own complex cognitive structures we can understand the way in which the spatial and other cognitive equipment of various animals approximate, in however imperfect a way, to devices more elaborately and complexly developed in ourselves; and we can suppose that we and these animals have evolved in our diverse ways while coping with a common environment. (Bartley, 1983, p. 863).

However,

> ... Modern science, physics, physiology, and psychology give one **finer** structures **from whose standpoint one can even criticize and evaluate one's own cognitive structures,** and identify and correct for distortions and limitations in them. (Bartley, 1983, p. 863).

The emergence of the human mind (including its self-awareness, reason, and capacity for objective knowledge) is an evolutionary novelty. This view of evolutionary epistemology goes beyond physicalism and mentalism: Our mind, though dependent on biological systems, is a new dimension of evolution. Therefore, although mental phenomena are evolutionary novelties, evolutionary epistemology does not amount to ontological reduction, but rather to **emergentism.** We argue that patterns of interaction on the organic level - that is to say on the level of brain cells - lead to the emergence of mental phenomena. But there is nothing supernatural about elements joining up to form a cycle and thus producing a new system whose functional properties differ fundamentally from the properties of all preceding systems (Lorenz, 1973). That means that the mind is not in the brain but that **it represents a specific functional property of the brain.**

The emergence of human mind is hence the beginning of a new stage of evolution. This new stage is the development of cultures, of objective knowledge systems, of scientific and philosophical theories. In short, it is the evolution of the products of the human mind (**world 3** in Popper's (1972) sense). And this is man's great opportunity! Our innate teaching mechanisms have indeed set bounds to our development as a biological species. As a **rational** animal, however, we can progress to a new level of cultural and spiritual evolution and thus transcend our biological bounds. To transcend the constraints of biological evolution also means to learn from evolution, to grasp these constraints realistically and even to overcome them by the means of reason. To understand human evolution, both biological and cultural, then, implies the opportunity to master our own future. It implies that for the first time in the history of our planet a living being has the chance to understand itself and to reflect upon its own future development.

V. EVOLUTION AND THE OPEN SOCIETY

Let me now formulate three consequences of the systems-theoretic, non-deterministic approach to evolution. These consequences might show the compatibility of the notion of the open society with biology and evolution theory.

The first consequence concerns human freedom. The systems view of evolution goes beyond the traditional alternatives of "determinism or indeterminism". In the course of evolution a series of novelties, among others the "invention" of the central nervous system have led to completely new patterns of organization which have preserved certain degrees of freedom. Since evolution is not pre-determined by rigid laws, it is, in a way, accessible to new dynamic regimes and, thus, to freedom; it "preserves its accessibility to freedom on every level of organic complexity" (Riedl, 1977, p. 367). In other words: evolution is an open program.

The second consequence, following from the first, concerns the structure of an open society. An open society, in contradistinction to a closed one, is a "society in which individuals are confronted with personal decisions" (Popper, 1962, vol. 1, p. 173). This definition agrees with a systems model of society outlined by Bunge that goes beyond individualism and holism. From the point of view of "systemism" Bunge (1979b, p. 16) writes that "a society is neither a mere aggregate of individuals nor a supraindividual entity: it is a system of interconnected individuals". Accordingly any interaction between two societies is an individual-to-individual affair, and any social change, is therefore, a change at both the individual and societal levels. I suppose that such a systemic view of society excludes the notion of a closed society and preserves the notion of the individual's accessibility to freedom and, thus, to personal decisions. Individual behaviour depends of course to a certain extent on the individual's genetic make-up and is influenced furthermore by social constraints.[8] But this does not mean total determination: There must be - and certainly there is - the category of personality common to human individuals. This category is missing only in totalitarian systems such as Plato's vision of the state or the Naziregime.

The third consequence follows from the previous two and concerns decision making and planning in the life of human systems. Undoubtedly, this is one of humanity's central problems. If we apply the concept of self-organization to society, we should take cognizance of the correlation between internal and external factors which characterize the behaviour of humans. Jantsch (1975) uses a simple diagram (Fig. 2) to make visible

> ... that human **inventiveness** alone can fulfill some of the tasks of operational planning in the realm of **purposive systems** - namely the invention of new operational targets. **Adaptive** systems may only select from a given spectrum of targets but have considerable flexibility in doing so since internal organizational changes may accompany the selection of such targets. **Mechanistic systems** are restricted in the same task by the rigid organizational structure which has to fit all selected targets. (Jantsch, 1975, p. 71; my emphasis).

This third consequence of the systems approach to the study of human social systems requires some further remarks.

According to Fig. 2 we may distinguish three levels of organization: (i) the **organic** level; (ii) the **behavioural** level (which is a systems property of organic structures); and (iii) the **rational** level (which is a systems property of man's behavioural organization).

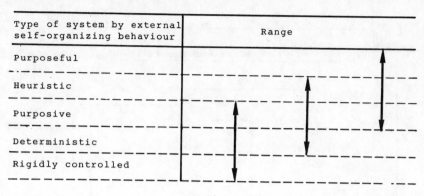

Type of system by external self-organizing behaviour	Range
Purposeful	
Heuristic	
Purposive	
Deterministic	
Rigidly controlled	

Mechanistic   Adaptive   Inventive

Type of system by internal self-organizing
behaviour

Fig. 2

The invention of new operational targets depends upon a rationality which
is an outcome of evolution and which is a specific property of the human
brain. However, the organization of the brain or the central nervous
system respectively represents an internal factor in human actions. A
second factor is given by social and cultural constraints and is, then, an
external factor.

I have tried to illustrate the interdependence between internal
(endosomatic) and external (exosomatic) factors in the structure of
individual actions and decision making in Fig. 3.

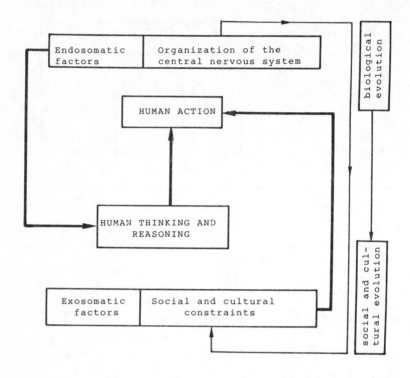

Fig. 3

Model of interactions between endosomatic (biological) and exosomatic (social and cultural) factors of human action. Thinking and reasoning are systems properities of the human central nervous system; besides these abilities are influenced by social and cultural circumstances. Thus human action always is a result of both biological and social (cultural) constraints. The central nervous system represents the biological basis of social and cultural behaviour; biological evolution has been the precondition of social and cultural evolution.

Human actions, in general, may be described by a model similar to **trial-and-error** models in biological evolution. Any successful action is performed again and again, in what certainly involves **habitualization.** Berger and Luckmann in **The Social Construction of Reality** (1966) point to the role of habitualization in human actions:

> All human activity is subject to habitualization. Any action that is repeated frequently becomes cast into pattern which can then be reproduced with an economy of effort and which, **ipso facto**, is apprehended by its performer **as** that pattern. Habitualization further implies that the action in question may be performed again in the future in the same manner and with the same economical effort ... Even the solitary individual on the proverbial desert island habitualizes his activity. (Berger and Luckmann, 1966, p.71)

The economical advantage of habitualization is self-evident. It narrows the many ways to go about a certain project to one and so "frees the individual from the burden of 'all those decisions', providing a psychological relief that has its basis in man's undirected instinctual structure" (Berger and Luckmann, 1966, p. 71).

But even more important than habitualization are man's **creativity** and **inventiveness.** Habitualization alone would lead to rigid structures of behaviour that would not allow for rational progress through the inventing of new solutions, that is to say new ways of acting, new ways to reach certain goals.

It will be easily recognized that the systems view reveals ambiguity in the development of man and society. On the one hand there are evolutionary biological constraints cementing certain pathways. On the other hand, however, an increasing complexity of systems has preserved the accessibility to freedom. Human thought and reasoning depend upon the structures and functions of the brain. Yet, the human brain has loopholes, so to speak, that preserve the feasibility of creating new ideas. Thus, humanity is - and always must be - open to new directions of thinking, using new arguments, learning from "old" mistakes, and so forth. Closed societies do not allow new arguments; only open societies do, and by so doing increase our flexibility and our chances for survival. "Either we learn from our biological and cultural history, or those who do not learn from it are forced to repeat endlessly the mistakes of the previous periods" (Leinfellner, 1984, p. 275).

The creation of new ideas, then, is man's great opportunity. But any idea taken absolutely can be a great peril. Bertrand Russell in his **Unpopular Essays** says:

> I think that the evils that men inflict on each other, and by reflection upon themselves, have their main source on evil passions rather than in ideas or beliefs. But ideas and principles that do harm are, as a rule, though not always, cloaks for evil passions. (Russell, 1968, p. 161).

Ideas that harm mankind very often have indeed been cloaks for man's evil passions. Moreover, ideas have again and again been the roots of ideologies. In the pursuit of truth men have often been misdirected by illusory styles of thinking (cf. Topitsch, 1979). In the pursuit of happiness men have tried to rescue "Utopia"; but instead of learning from their own history - and, thus, from their mistakes - they have instead repeatedly created prophetic arguments. "The prophetic argument is", as Popper (1962, vol. 2, p. 156) explained, "untenable, and irreparable, in all its interpretations, whether radical or moderate". Furthermore, any prophetic argument is contradicted by **objective knowledge.** The search for objective knowledge is in a way, an expression of the open society. Political leaders of closed societies rarely allow objectivity in knowledge for they try to substantiate their own ideologies. To deny or even to reject the idea of objective knowledge then would have bad consequences for philosophy and for society. (On objective knowledge in science and philosophy see **e.g.** Popper, 1972; Radnitzky, 1980; Radnitzky and Andersson, 1978; Riedl, 1980).

Finally some words about **ethical** demands in the context of biology and evolutionary theory. Let me say at once that from the viewpoint of evolutionary epistemology our moral behaviour is also a result of evolution. Such a statement does not contradict the classical distinction between **is** and **ought.** Moreover, it has become very important today to distinguish between **evolutionary** and **evolved ethics** (Tennant, 1983), because of the danger that the sociobiologists' metaphor of the "morality of the gene" (Ruse, 1984c) might be abused by awful ideologies. Morality has evolved during evolution, but evolution does not - and cannot - compel ethical decisions from us. From the non-determinist viewpoint only the capability of moral behaviour is a given category. There is no law of evolution to prescribe how we have to fulfill moral demands.

However, there is one imperative of greatest importance for man's survival: to try to "harmonize" with nature. Julian Huxley's **evolutionary humanism** (1957, 1959) is of some consequence to our reflections on our place in the universe. During our cultural evolution and particularly as we have applied technology we have changed nature to a great extent. More than that: We have destroyed many natural systems - and now run the risk of destroying ourselves! This has at least two causes.

First, many thought systems, many ideas which have been created about nature and the place of man in nature are the products of **idealistic** philosophy. Idealistic philosophy consists of a bulk of unworldly interpretations of the world. It is, as Popper (1972, p. 32) emphasizes, "the greatest scandal of philosophy ... that, while all around us the world of nature perishes ... philosophers continue to talk ... about the question of whether this world exists". By means of evolutionary theory and evolutionary epistemology - in opposition to idealism - the world, or at least parts of it, can be grasped **realistically.**

Second, up to now we have obviously not been willing to learn from our own biological and cultural history. What we can in fact learn is how to plan the future for our own survival. In fig. 4 I have tried to skeletonize the basic ideas of an "open evolution" based on systems conditions.

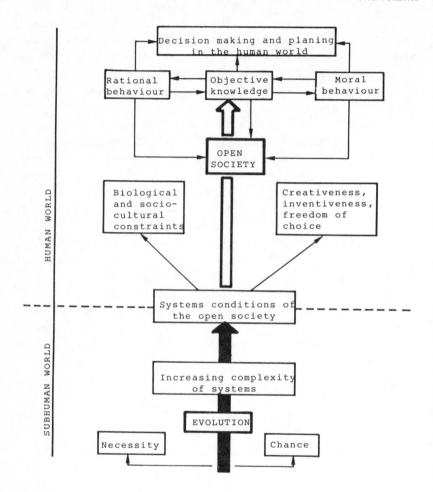

Fig. 4

The diagram aims to show biological evolution as a process combining chance and necessity: Increasing complexity of systems led to the emergence of man and human culture and society. Thus, on the one hand there are biological as well as socio-cultural constraints. On the other, human creativity and inventiveness, by applying new ideas, can always transcend these constraints. Biological evolution has not determined socio-cultural evolution: it only has made the emergence of complex and highly organized social and cultural systems possible. Hence an open society would be one in which each individual is given the chance to create and to invent new ideas which might be of importance not only to himself but to the public welfare as well. Furthermore, decision making and planning in an open society are based on three premises: rationality, objectivity and moral judgement.

VI. CONCLUSION

I feel that I have raised more problems and questions than I am able to answer. However, the goal of the present paper has been to stimulate further discussions at the interface between biology and the human sciences.

So I hope that I have hinted at the compatibility of modern evolutionary theory with the notion of the open society in a somewhat fresh manner. I want to stress once more that biology and particularly the systems approach to evolution is not compatible with prophetic arguments and that it does not accord with any kind of determinism (which would be a dangerous ideology). Man himself is the maker of his world - and, therefore, his development as an **animal rationale** rests with him.

From a biological point of view there is no reason to worry whether evolution will justify us. We do, however, have good reason to worry whether we will be able to justify our behaviour, our way of acting **in** - and **against** nature.

NOTES

*   I presented parts of this paper at the "First Meeting of the Open Society and Its Friends" in New York City, November 22-24, 1982. (This Society was established on the occasion of Karl Popper's 80th birthday). However, I have rewritten the original manuscript in order to adapt it to the targets of the present volume. As the reader will recognize, some of the basic ideas of this article are strongly influenced by the work of Popper. However, regarding many other problems - **e.g.** the mind-body problem - which are not explicitly discussed in the present paper, I do not agree with Popper's solutions.

1.   Popper's public discussion of biological problems is comparatively recent. According to Bartley (1976) Popper started this discussion in the afternoon of Tuesday, November 15, 1960. On that day Popper, in his seminar, read a new paper of his own, which was mainly concerned with biology and the theory of the three worlds. Later on Popper developed his view of evolutionary epistemology in accordance with modern biology (see Campbell, 1974).

2.   In the following review of problems and approaches in evolutionary theory I must take a shortcut, going straight to the central questions of the present paper. I have discussed the systems view of evolution in addition to synthetic theory in some other books and papers (see my **Grundriß der Evolutionstheorie**, 1982b). My approach to evolution has been strongly influenced by Rupert Riedl and supported by Günter P. Wagner. The latter was, like myself, Riedl's student at the University of Vienna in 1973 - 1978.

3.   The systems view in biology was established in the 20' by Paul A. Weiss and Ludwig von Bertalanffy. I trust I am not doing Weiss any injustice by saying that Bertalanffy's contribution to systems theory has been the greater extending general systems theory to a kind of new natural philosophy (see Bertalanffy, **General System Theory**, 1973). Incidentally Bertalanffy was a colleague of Konrad Lorenz and a teacher of Rupert Riedl at the University of Vienna during 1946 - 1948. It may be justified, therefore, to speak of the "Vienna School of Theoretical Biology" which has its roots in the late 40' and which continues up to now.

4.   Gould, however, like some other paleontologists (**e.g.** Niles Eldredge, Steven M. Stanley, and others) advocates the **punctuational** model of evolution which in some ways resembles the "saltationist" argument.

5.   The term "teleonomy" was coined by the ethnologist C. Pittendrigh. He intended for this term to describe goal-directed activities in living systems in a less obscure terminology, supplementing the concept of teleology which had been preoccupied with metaphysics or at least by philosophical schools closed to metaphysics.

6.   A worthy attempt to understanding evolution and, particularly, the emergence of order in living organisms comes from **nonequilibrium thermodynamics**. This discipline could bridge the gap between physics and biology, between the traditionally closed systems of the former and the open systems of the latter.

7.   "Mesocosm" means that section of the world to which our cognitive apparatus (and the respective cognitive apparatus of other organisms) has been adapted. It is, so to speak, a world of medium dimensions. Vollmer (1984, p. 87) writes: "By analogy with the evolutionist's concept of an 'ecological niche' we could call that section a 'cognitive niche'... Every organism has its own cognitive niche or ambient, and so does man".

8.   During the last ten years **sociobiology** has rapidly developed as a discipline which aims at giving genetical explanations of social behaviour. I have objections against the claim of sociobiologists that **human** social behaviour can be reduced to genetic mechanisms all along the

line. But I shall refrain here from specific criticism of sociobiology and the sociobiology debate. (For details see Ruse, this volume). I only want to mention one book **pro** sociobiology (Wilson, 1978) and one **against** sociobiology (Lewontin **et al.**, 1984). I also draw attention to a philosophical analysis of sociobiology by Ruse (1984a) and a critical evaluation of the sociobiologists' claims from the viewpoint of a social scientist (Meyer, 1982; this volume). I am afraid, however, that at least some sociobiologists - whether consciously or unconsciously - are going to support ideological doctrines.

MICHAEL SCHMID

COLLECTIVE ACTION AND THE SELECTION OF RULES.
SOME NOTES ON THE EVOLUTIONARY PARADIGM IN SOCIAL THEORY

The purpose of this article is to give a brief and hopefully concise summary of the importance and characteristics of the role played by evolutionary thought in the discussion of social theory. In order not to lose the thread of my argument in the multiplicity of theoretical and meta-theoretical considerations which have arisen in the course of this long-standing discussion, I have decided to concentrate on four points.

(1) First, I shall attempt to analyse the genesis of the theory of evolution, insofar as this genesis is relevant to the problems encountered in the development of social theory.

(2) Next I shall reconstruct the central argument of a theory of evolution, and

(3) try to show its importance in the development of a general social theory of action.

(4) I shall conclude with a discussion of some of those exemplary explanations which many social theorists apply in the conviction that they will contribute to the accuracy of their evolutionary core-assumptions.

I. ON THE GENESIS OF THE SOCIAL THEORY OF EVOLUTION

It is a popular notion that social theory as a scientific enterprise is the product of the events witnessed at the end of the 18th and the beginning of the 19th centuries when it was commonly realised that social structures, i. e. the traditional norms of social conduct, broke down and disintegrated as the result of industrialisation and its after-effects (Poggi, 1972). Reactions to these events took various forms, the attempt to explain the changes in society in terms of evolutionary theory without lapsing into metaphysical or mere historiographical interpretations being

M. Schmid and F. M. Wuketits (eds.), Evolutionary Theory in Social Science, 79–100.
© *1987 by D. Reidel Publishing Company.*

but one of many (Mandelbaum, 1971; Hawthorne, 1976). For social theorists, however, who wished their theoretical ideals to keep pace with the successes of the so-called inductive and generalizing natural sciences, the idea of social evolution proved to be forward-looking insofar as it enabled them to understand and systematize social developments with the help of general "evolutionary principles"[1]. Unfortunately it soon became evident that, despite this imitation of Newton's metatheoretical ideas, the uniform and paradigmatic theoretical position desired by so many remained an unfulfilled hope. It turned out that the seemingly uniform concept of social evolution embraced two opposing, indeed incompatible views of the course of social events.

The **Spencerian** variant of evolutionary thought (Nisbet, 1969, pp. 166-188; Peel, 1971, pp. 131-165)[2] emphasized the necessity of distinguishing between different levels or stages of development of societies. This process it was hoped, could be empirically related to actual historical forms of societal reproduction. It was at the same time intended to detect at least inductively the basic laws of progressive evolution, which made the increasing complexities of communal bonds and the advances in the economy as conceivable as the emergence of modern political, domestic and judical institutions. The hope here was to understand this evolution of institutions as a kind of "development" or "unfolding" of endogenous potentials enabling primitive or archaic societies irreversibly, necessarily and teleologically to reach what is now known as the "modern age". Unfortunately however, this hopeful programme remained unrealized since scientists could not in fact identify any so-called "laws" or "principles of social evolution". Consequently the empirically identifiable transitions were left without a true theoretical basis, and in the end rested on assumptions which amounted to no more than a repetition of the original empirical discoveries, according to which human societies had developed from their formerly homogenous, relatively unstructured forms into more complex and highly-structured ones. Thus the Spencerian programme remained mainly descriptive (Turner, 1985, pp. 85-106) and thus restricted an apparently philatelic construction of social typologies. From these typologies it proved very difficult to draw any theoretical conclusions, because even very clearly defined structural categories could not provide valuable insight into those underlying

causal mechanisms which were the driving force of the postulated
development from simple to more complex forms. This grave and even
damaging insufficiency was further intensified by the fact that (to this
day) there was no common agreement on a workable criterion permitting a
clear-cut construction of different levels of societal complexity
(Parsons, 1967, pp. 460-520; Parsons, 1977a, pp. 230-241; Granovetter,
1979, pp. 489-515; Abercrombie, 1972, pp. 47-51).

Considering the confusion which sprang from this state of theoretical
affairs it is hardly surprising that despite the all-embracing claims the
Spencerian programme made for   explaning the total development of human
societies, support gradually waned and an alternative programme, that of
Charles Darwin, began to supersede it. Darwin's theory of natural
selection had   one singular   advantage over   Spencer's: It did not set
out to explain the actual course of history in all its details and global
directions, including the undeniably increasing complexities of biological
organisms. Rather it was content to investigate the central causal
mechanism which could explain just **why** certain morphological structures or
patterns of behaviour were retained in certain populations while others
underwent long-term changes or even extinctions. Only then did the theory
of evolution take shape as a valuable explanatory programme of research
(Ruse, 1979, p. 270).

But the transfer of Darwin's theory to social theory was destined to
meet with reservations. Unlike Spencer, Darwin intended to restrict
evolutionary theory to biology and seemed very reluctant to accept any
"creative force of evolutionary principles" behind the course of socio-
historical events. Thus it was only with some hesitancy and without open
acknowledgment that he took note of Marx's suggestion that it was in no
way unreasonable to interpret Marx's theory of the rise and fall of
various modes of societal production within the categories of a theory of
natural selection (Hoyer, 1982, pp. 15-23). In addition a racist
interpretation of the concept of "natural selection" had a discrediting
effect ideologically, on the application of Darwin's ideas to the social
sciences (Francis, 1981, pp. 209-228; Hofstädter, 1969, pp. 170-200).
Moreover, other, more logical misgivings came to the fore. Despite
repeated indications that the central line of thought in Darwin's theory
had been borrowed from the market theory and the demographic theory of

Adam Smith and Thomas Robert Malthus respectively (Hayek, 1969, pp. 127, 156-157; Hayek, 1973, pp. 22-24; Darwin, 1982, p. 93), any attempt to argue the case for the theory of selection in the field of social theory was inevitably accused of falling prey to the doubtful use of analogy (Stebbins, 1967, pp. 223-234). To quote just the most popular objections to the indiscriminate transference of biological arguments to social events it was quite evident that social systems reproduce not biologically, as natural species do, but by cultural and social learning, by establishing oral and literary traditions, by institution-building etc. all of which have no parallel in the bio-sciences. Certainly human behaviour did not seem to be determined by biological drives or instincts but to be regulated by changeable and debatable rules of conduct. And nobody should plausibly argue that societies "die" in the same way as biological organisms do.

But justified as such warnings may be, and even giving them their due, they cannot fully discredit the application of Darwin's theory to the social sciences simply because they overlook one important factor. The theory's potential fertility in no way depends on whether or not it can act as a backdrop for drawing analogies between biology and social theory. It depends rather on the more important fact that it nurtures, I propose, a legitimate hope that the selection-theoretical arguments will be justified in both instances, if one bears in mind that both Darwinian (or Neo-Darwinian) biology and the theoretical social sciences, are equivalent, albeit different examples of the use of **one and the same general theoretical calculus (or model),** the logical structure of which remains the same. In other words we should not expect that we could ever transfer the **specific** processes from one field of application to the other with recognizable theoretical gains. All we can envisage is to understand them in each case as **selective** and **structural processes.**

In order to strengthen this position I intend to give a brief reconstruction of the logical "hard core" of a theory of structural selection by enumerating (at least four) essential and necessary conditions which seem enough to form the basis of the theory's most general use. I will then go on to argue that the central social theory, the theory of action, fulfills these criteria, whereby it becomes admissible to understand it as a hopeful example or paradigm of the general theory of selection.

II. THE LOGICAL STRUCTURE OF A THEORY OF STRUCTURAL SELECTION

My explication of the logical character of a selection-theoretical calculus is based on the following conditions (Giesen and Schmid, 1975; Giesen, 1980, pp. 56-61; Giesen and Lau, 1981; Lau, 1981, pp. 44-48):

(1) One must be able to identify a population of individuals (no matter what the level of structural emergence or what the actual combination of their attributes) who can reproduce[3]. The extent to which this proves to be possible defines the size of that population. "Reproduction" here means, generally speaking, the transmission of a programme or code (a system of generating rules) on to the next generation, who because of this reproductive link will show a visible likeness to their predecessors.

(2) Subsequently, the assumption should be accepted that this reproduction is not always achieved without some loss of information or without programmatic changes. In other words: During the process of transmission this "copying" is, if you like, regularily subject to some errors or mistakes, either because parts of the programme become interchanged or, more generally speaking, recombined, or because completely new elements sometimes find their way into the copied programme through mutations, diffusions, inventions, or whatever. Therefore every successful reproduction of a population finds itself permanently faced with the problem of so-called "variation".

(3) Accepting these two conditions it should be always possible to differentiate, categorially and empirically, between the programme to be transmitted and the overt characteristics of any member of the population which it directs and determines. This is an important point since I choose to agree with the view that environmental factors do not have any direct influence on the programme in question, but only an indirect one affecting either the overt characteristics or the behaviour of the individuals and the consequences thereof.

(4) With this the fourth and final condition has already been touched upon. One should not regard the environments of a reproductively capable population simply as free scope, as a "playing ground" for the adaptive potentials of that population, but rather as a selective ·factor, which helps determine causally just which parts of the programme can reproduce

and which cannot. Basically populations find themselves under pressure of natural selection because of their environments. They cannot escape at will, however much they may wish to; the environment forces them to adapt. Later I shall argue that these environmental pressures can be qualified in terms of scarce, but nevertheless indispensible resources and their respective allocation.

On the basis of these conditions it is possible to formulate some abstract, albeit far-reaching hypothetical assumptions (Schmid, 1981; 1982a): Obviously, if there is a consistent adaptive relationship between a certain environment and a population, capable of reproducing within it, then a constant, **stabilizing process of selection** can be relied upon. We may in other words suppose that a definable optimum (of reproduction) has already been achieved consequent to the fact that such a process eliminates any non-adaptive variant. Any variant, whatever the cause of its existence in the programe, which disturbs the once established equilibrium will not be passed on with any degree of permanence (or at least only as recessive information).

Should the environment change in a way which affects the optimal reproduction negatively, we can say that the respective population envisages an additional environmental pressure. In this case those contingent variations, already in existence, which help achieve and retain a new local equilibrium, will be transmitted with increasing probability. In such instances we may speak of a **directed selection**, the course of which (and we must be quite certain of ⸀this) rests as well on corresponding environmental changes and not exclusively on the endogenous standards of the individuals concerned.

If the environment diversifies in a heterogenous way and several different niches thereby emerge in the immediate vicinitiy, then the accumulation of mutative or adaptive changes in the transmitted programme can have **diversifying** or **speciating** consequences. The end result is the emergence of different populations no longer capable of reproducing with each other, in a word, species.

These three kinds of selective processes form a coherent whole. Together they allow us to explain within one and the same theoretical framework deriving from a finite set of variables, i. e. reproduction, variation and selective retention, and an equally finite set of functions which realates them to each other, why certain groups or populations change or remain as they are. These variables and functions combined into a closed and interconnected theoretical system provide the "hard core" (Lakatos, 1970, pp. 132-138) of a wide and abundant research programme, which, because of the need to explicate and qualify the postulated functional relationships between these given factors, establishes an extensive heuristics, which can in turn aid and direct further theoretical and empirical research.

III. AN ACTION-THEORETICAL INTERPRETATION OF THE THEORY OF STRUCTURAL SELECTION

In order to judge the potential productiveness of such an abstract model or calculus of differential selection for the study of social sciences I shall have to examine whether the four necessary conditions mentioned above can be fulfilled in its area. To do so I have to accept a somewhat presumptious argument in contradiction to widely held opinion. I do not agree with the idea that social theory (encompassing contributions from quite different academic branches like sociology, political science, economics, anthropology, history and so forth, each with all their corresponding internal differentiations) is a kind of "multiple-paradigm" discipline (Ritzer, 1975; Eisenstadt and Curelaru, 1976), which embraces various, heterogenous and in part incompatible strands of thought. Instead of this idea of a frayed "theoretical patchwork" I am inclined to support the ideal of a unified or standard tradition of theorizing in social research, able to integrate all those different approaches. I am thereby continuing in the tradition of a comprehensive programme of research defended by Talcott Parsons, on the basis of the work of a series of "classical" sociologists and economists such as Emile Durkheim, Max Weber and Vilfredo Pareto as a unified action-theoretical paradigm (Parsons, 1968[2], 2 vols.). According to this paradigm any social theory necessarily and exclusively deals with actors, their actions and the social situations which emerge in consequence of the very fact that actors have to act in relation to each other. The present-day academic division of labour between different social scientific disciplines as it irrefutably exists can be theoretically accepted only presuming that each of those diversified theoretical approaches at least in principle can justify itself as a (logical) part of this comprehensive action-theoretical paradigm.[4]

Such a paradigm can be introduced in two stages. First of all it is based on the following (simplified) assumptions (Parsons 1968[2], vol. 1, pp. 43-86):

(1) **Only persons act.**

(2) Such persons are not regarded as (biological) organisms, but as **actors** who, have (among other things) the capacity of choosing between different, more or less preferable **aims.**

(3) The actors are therefore in the position within the confines of specific cognitive, emotional and motivational prerequisites, to select actions as **means** of achieving their aims.

(4) They do this in **objective situations,** the characteristics of which, at least in regard to the actions concerned, cannot be changed nor influenced by the actors and to which they consequently have to adapt.

(5) An actor can orderly rank those aims open to him and choose the best possible means of realizing them only if he can rely on **normative standards** (or so-called "normative orientations") which justify his choices of ends and means.

According to Parsons these assumptions help designate the basic variables of a **theory of individual action** for which appropriate (theoretical) functions must be found to link them. Evidently such a theory has only restricted explanatory power; it deals exclusively with the individual actions of single actors. Consequently we need to extend and complete it, for it is only by applying this theory to plural actors that we come to the core problem of an action-theoretical research programme. The problem is one of ensuring that actors who have free choice over the means and ends open to them can cooperate on a permanent basis, without running the risk that all their attempts to reconcile their various actions will have destructive consequences. Or to put it in other words: Under what conditions can we ensure an ordered, calculable and common course of action between all the actors concerned in view of the fact that the social situation of each individual is determined quite emphatically by the very existence of the other actors, all of whom find themselves in exactly the same circumstances?

The task of answering this question has fallen to an autonomous **theory of collective action.** This theory is based on the following assumptions (Weber, 1964, pp. 16-41, Schmid, 1982a, pp. 112-128):

(1) If social or collective action is to succeed repeatedly, then certain **rules** must be observed. No permanent collective action is possible without a common set of rules. Failure to agree upon and to institutionalize binding standards of social behaviour may be taken as an indication of a state of anomie in a particular action system.

(2) Such rules for social action, or to be more precise, **rights** and **norms**, stipulate a code, a central core of standards or a constitutional programme (Vanberg, 1982; 1983). This is, in short, a "deep structure" towards which the actors orientate themselves in social matters. This does not necessarily mean that they follow these rules, only that they take them into consideration in preparing their actions. Regulations only gain actual validity when this is the case.

(3) Any observance of these rules by more than one actor (**i. e.** in social situations) inevitably leads to some collective consequences of their actions, to so-called "distributive structures" in the sense of the macro-social theory of Blau (1977) and Giesen (1980, pp. 41-45). Thus the competitiveness of a market economy unavoidably results in an unequal distribution of those possessions and opportunities which determine the further course of market transactions by any of the participants (Parsons, 1977b, p. 334; Merton, 1964, pp. 131-160). The main theoretical problem in this context is necessarily whether or not, and if so to what extent, such consequences of collective actions back up the set of established rules from which those actions originally derived.

(4) The core of a theory of collective action is the assumption that specific processes can be made responsible for this kind of rule-stabilization or rule-destruction, either of which may be described as a "process of selection". These processes of selection (some details of which I shall discuss later) determine just how far the actors are likely to continue to orientate themselves towards the existing codes, and in doing so to "co-ordinate", in a general sense, their behaviour with that of their co-actors or wether in consequence of their failure to coordinate their mutual behaviour the rule set is changed or even driven to extinction.

Two factors must be considered in estimating the likelihood of
varying rules and regulations being maintained:

(1) First, it must be made quite clear in each case exactly what the
collective consequences of common actions are when people observe certain
rules of conduct in social situations (or do not). In identifying these
consequences it is irrelevant to guess whether they are intended, expected
or desired. It is more important to see that the scope of action available
to the actors is objectively determined by these consequences. In other
words, their decision as to whether or not they continue to comply with
long-standing codes of conduct is not exclusively affected by their
definitions of the collective situation (i.e. by their cultural concepts
and cognitive knowledge) but rather by the opportunities and restrictions
which result form the very fact, that those collective concequences arise.
If envisaging these restricted opportunities the actors are able to keep
up their mutual cooperation, then their collective actions (or the
corresponding rules) can be called "functional", or "adaptive". If they
cannot, "social contradictions", "paradox effects", "dysfunctions" or
"structural tensions" occur[5]. When such conflicting consequences are
unavoidable then the system of rules causing them comes under selective
pressure.

Under these circumstances it becomes more likely that innovations or
variations of the existing rules will be accepted to the degree that they
objectively help prevent those collective effects which violate the former
rules and if, in addition, they create new, self-supporting consequences
of their own. Of course, one cannot ensure that the acceptance or institutionalization of new rules will not in themselves have even more
damaging consequences. On the contrary, one should always reckon with
this. It is exactly this inevitability of collective consequences acting
to destabilize rules of social action which sustains any further evolution
as regards to rules and regulations. If one concentrates one's
theoretical analyses on the possible effect of collective actions for the
further reproductive capability of those rules one should speak of the
problem of "internal selection".

(2) But systems of rules do not evolve exclusively as a result of
consequences of collective and rule-regulated actions; their evolution is
equally influenced by the demands for external resources which cannot be

obtained within those regulated systems. The specific distribution and allocation of resources determine the environment of systems, a set of resource-defined niches to which systems have to adapt because their control never can be complete. Codes of conduct are therefore inevitably exposed to the selective influence of these external exigencies. Innovative proposals for new rules consequently have but a slim chance of gaining collective recognition not only because they might endanger existing and successful institutions but also because of the selective effects of differentiating environmental changes in the distribution of those resources which are necessary for the reproduction of these institutional rules. This selective impact of resources on the stability of regulations can be described as "external selection". Internal and external selections are factually connected if the consequences of collective and rule-guided actions result in a critical shrinkage or even the final exhaustion of resources.

Combining these lines of thought we can now propose the central argument of the theory of collective action with greater precision. Both the continued stability of existing codes of conduct **and** their changes (including the stability and changes of the corresponding collective actions) are dependent upon which contingently existing rules are likely to be given preference over readily available alternatives, in view of the specific collective-action consequences which in turn result from those rules, **and** regarding the differential pressure arising from a specific distribution of externally localized resources. What form and direction the structural changes take depends on the kind of collective consequences of actions which emerge in the course of established processes of selection and on the availability of reproductively important resources irrespective of the fact whether these resource-distributions are dependent on those actions or not. We thus suppose that both stability and change of regulative structures are the result of the impact of a dynamics of **differential selection.**

Even if one accepts such a reconstruction of the central action-theoretical research-programme in terms of the theory of selection, the question still remains whether the programme in fact satisfies the four necessary conditions stated at the outset to the extent that it can be considered a valid case of application of that general calculus of selection. I believe that this is indeed the case[6].

(1) Actions which comply with specific sets or subsets of regulations and which in so doing gain their specific "meaning" or "sense" (Winch, 1958; Luhmann, 1971a, pp. 25-100; Luhmann 1984, pp. 92-146), can readily be interpreted as populations of actions. The form and content of the rules which actors might use to construct their corresponding actions determine and confine the scope or extension of the corresponding population of actions. The first prerequisite is thereby fulfilled; considering actions and the rules that generate them as populations of actions may prove useful.

(2) Furthermore, such populations of actions possess the necessary qualification of being reproductive. Patterns of action can be transferred in so far as the actors take over from and hand over to each other the rules which they might observe (using imitation, socialization or any other possible mechanism). Mutual understanding of proposed regulations and, their eventual acceptance are of course indispensible for their successful reproduction. One can easily see that because of this stipulation a smooth transmission of such rules cannot always be expected: rules can be misinterpreted or deliberately ignored, they can be combined differently, or new, fomerly unknown rules can be thrown open to discussion. That is, deviations and variations of rules must, of necessity, **continually** arise (Lamnek, 1979; Roberts, 1979; Luhmann, 1972, 2 vols.).

(3) Sets or systems of rules define behavioural programmes or codes which must be activated or realized by specific actions (see point 1). Thus rules and actions and their subsequent collective results are to be kept strictly apart (Bourdieu, 1979; Waddington, 1975). Only then does it become plausible that certain rules are exposed to selective pressures which arise from some of the more averse consequences of the very actions they generate. Thereby the third of the above mentioned criteria has been fulfilled.

(4) The economic theory of the firm, or more generally speaking, some of the recent theories of organizations rest expressly on the view that social structures or regulations rely on external resources which cannot be obtained internally. The scope and kind of actions which are open to the actors affected by this restriction are thus determined externally (Aldrich, 1979; Nelson and Winter, 1982; Hannan and Freeman, 1976). Hence

it follows that external resources play a decisive part in the selection of those courses of action which might help to solve the urgent problem of their actual acquisition. Correspondingly only those variants of rules which generate actions capable of successfully acquiring these needed resources can be of any reproductive permanence. Our reconstruction of a selectionist social research programme encompasses this view and thereby satisfies the fourth prerequisite of a calculus of selection, which unambiguously emphasizes external circumstances as active, causal selectors of internal structures.

To sum up: The social theory of action satisfies the stated conditions of the calculus of selection and can thus be regarded as one of its promising paradigms.

This concludes my resume of the central social theory of action. In the final section of this paper I shall proceed to some widely discussed examples which may support the productiveness of such a research programme.

IV. THE HEURISTICS OF THE THEORY OF STRUCTURAL SELECTION

For centuries social scientists have been preoccupied with two general problems: First: How can we realize a just social order (or social structure of mutual relations) one which is understood to be based on a set of commonly shared rules of conduct, without being forced to accept unjustifiable and undesirable collective consequences of those interrelated actions which might endanger the demands and needs of the members of a society? (Elster, 1978; Ullmann-Margalit, 1977; Opp, 1983; Schotter, 1981; Parsons, 1968[2], vol. 1, pp. 87-125.) Second: How can long-range changes in social structure be accounted for?[7]

A number of the answers frequently given to the first of these questions seem to remain valid to this day. In the traditional line of social thought three different mechanisms in particular are acknowledged to be responsible for the origin and duration of common rules.

(1) First, there is the legitimized use of power, i. e. **authority** (Weber, 1964; Wrong, 1979; Lasswell and Kaplan, 1950; Hennen and Prigge, 1977; Eder, 1976). The threat of authoritarian reinforcement of existing rules might prevent those less welcome collective consequences of common

actions which result from the fact that some actors, whatever their
reasons, permanently refuse to conform to common rules. Such actors take
the benefits of social order as a kind of public good, without being
prepared to share the costs of its establishment and legitimization.
Without any authoritarian supervision we may expect at least a fluctuating
minority of actors to assume that because everyone else is inclined to
abide by the rules they can further their own individual interests by
unilaterally bending and breaking them, thus injuring the interests of
those co-actors who decide not to do so. To avoid such harmful and
possibly anomic constellations people might be ready to accept some kind
of institutionalized power which by authoritarian penalization of deviant
actors exacts a price for any individual infringement of once established
rules, and thus stabilizes their actual validity. Theoretically speaking,
authoritarian institutions might be selectively favoured in strategic
situations where people may endanger the cooperative interests of all by
their rule-ignoring egoistic behaviour. In such notoriously unstable and
possibly anomic situations, where the social order is in danger forms of
authority present local attractors from which one can deviate only if
alternative institutions emerge that minimize (or do away with) the costs
which arise from an abolition or disintegration of recieved power
institutions. Inherent morality, i.e. feelings of honour and
selfcommitment, and mutual trust might be such functional alternatives.
They are usually covered by the term "solidarity". Unfortunately these
alternative procedures do not allow for a desirable stable equilibrium
(Ullmann-Margalit, 1977, pp. 36-37, 46-48).

(2) Another possible way of preventing or at least minimizing the
concomitant costs of establishing authoritative institutions (for instance
the unavoidable political inequality and the collective costs of criminal
justice and sanctions) is to hold **competitive processes** responsible for
the maintenance or the likely recurrence of desirable collective
consequences of common actions (Smith, 1976; Willecke, 1961; Hayek, 1976,
pp. 122-155). It is important to note that such competitive devices
possess self-sustaining and even self-accumulating characteristics which
might help us understand their functioning: If different actors accept a
set of rules of competition and are thus prepared to act accordingly as
they compete for scarce goods and resources, an unequal distribution of

these goods and resources amongst the competitors will inevitably result
over time. This inequality in turn defines the subsequent entry conditions
for continuing the competitive game as people are forced to start from
uneven lines of departure. As long as no countervailing factors intervene
we might easily expect that the once established unequal distribution
structure will be repeatedly re-enforced. Accepting these assumptions one
may conclude that a distribution permanently impairing the competitive
prerequisites of at least some actors might (at least under circumscribed
conditions) endanger their motivation to continue the contest in
accordance with the original rules[8]. They may in fact be readily
inclined to circumvent or openly break those rules in order to   score
gains by illegal means. Thus we should expect rules of competition to be
inherently unstable and in need of support by reinforcing supervisory
efforts. And indeed, if a certain critical number of participants cannot
refrain from striving for deviant and illegal advantages and if nobody
interferes by authoritarian means the competitive order dissolves and
becomes extinct. But as I have already indicated such interference has
its own collective costs and is often institutionally unrealizable. Thus
we may wonder why we observe competitive regulations at all. I believe
this can be explained.   First, we should accept as plausible the
theoretical insight that deviant behaviour only pays as long as a vast
majority of competitors observes the rules in question. Otherwise, if
nearly everybody uses illegal means to reach his goals practically nobody
can avoid the highly disagreeable situation of being a swindler being
deceived (Ullmann-Margalit, 1977, pp. 71-72). This collective situation is
much more unstable (and therefore rarely to be observed empirically) than
the above mentioned one where authoritative control and deterrence keep
the number of deviant actions small enough not to endanger the general
legitimacy of the rules concerned.

But there are two further reasons for the surprising stability of
rules of competition. First even if overtime an increasing number of
actors becomes inclined to alter established rules it is unlikely they
will succeed in agreeing on a feasable collective proposal to do so. In
anonymous situations, where decisions are made independently and
individually, and where actors do not establish permanent and personal
interactions, communications and control, there is no guarantee that each

of the other participants will accept any proposed rule changes. This is
especially true for those actors who profit by the distributive inequality
of a competitive institution and thus are prepared to protect it,
sometimes by any possible means.

The second reason is this: Competitive systems generate a progressive
divisions of labour resulting from actors' attempts to avoid being
excluded from those niches of reproduction which are more easily and more
profitably exploited. In reaction to this rather unstable and nasty
situation actors usually prefer to move to new and alternative fields of
action where competition is limited. As a result of this strategy a
growing interest in exchanging labour results and goods emerges which
leads to a richer network of mutual exchanges. The more extended this
division of labour becomes and the more extensive those networks are, the
less likely it will be that any individual participant can afford not to
take part in the transactions and the more stable the rules will be that
keep them alive. Emile Durkheim and Herbert Spencer have given a nice
subtle description of this development (Durkheim, 1964, pp. 101-132, 256-
282. Spencer, 1897, pp. 340-411; Corning, 1982).

Of course, competitive systems are not ultra-stable. There are
conditions which, if disturbed, discourage to regain a new state of
equilibrium. Extensive and oppressive inequalities in the distribution of
opportunities (Moore, 1978) endanger it as much as does the unavailability
of preconditions such as (exchange-)morality (Smith, 1977[2]; Weber, 1965)
or the rule of law and public goods (Smith, 1976, Book V), which as the
founder of the modern market-theory already knew, cannot be prepared by
those competitive processes themselves. But as long both deficiencies can
be avoided competitive systems can continue to function as effective
means of distributing goods and services in large and anonymous
populations while excluding any alternative rules of distribution.

(3) A third mechanism of selection seems to be less attractive: the
**formation of consensual agreement and commitment to a general
acknowledging of rules and values through discours**[9] Not only because
individuals, unrestrained by any form of authoritative protection could
too easily terminate any possible agreements not in their own interest,
but more so because the agreement and acknowledgement of common rules by
all participants is extremely time-consuming and usually can only be

achieved with any degree of success in small groups (Giesen, 1980, p. 38). In addition, if one is to believe Jürgen Habermas, such consensus is dependent on resources of well-meaningness, common sense and rational knowledge, that are not always prevalent (Habermas, 1971, pp. 101-141; 1981, vol. 1, pp. 369-452).

In many instances, of course, such regulations seem to meet with well established, unanimous agreement. On closer observation however it becomes obvious that even in these cases, agreement is reached through authoritarian pressure (or oktroy), through mere blind acceptance of the inevitable, or simply for the sake of tradition itself. This has been repeatedly pointed out by Max Weber (1964; pp. 20, 26-27, 157-159, 195-197). I feel that this pessimistic observation is true even today. And additionally one cannot discount the fact, that even in situations where consensual methods might successfully by introduced, they might nevertheless have damaging collective consequences which could weaken the institution of common and discoursive consensus-formation in favour of rather authoritative problem solving. Nevertheless, despite this inevitable insufficiency recourse to discourse remains the method of last resort in selecting rules, as social constitutions are always under the pressure of mutually incompatible interests which cannot be kept forever latent by authoritative or competitive devices. In other words: Social relationships where conflict-minimizing deliberations no longer have any possible success will inevitably break up, whether due to the migrative dissociation of the disadvantaged or, if this is impossible, to probable protest against the illegitimacy of the existing rules or constitutions (Hirschman, 1970).

Even if no ultimate answer has yet been found to the questions of whether and in what way these various mechanisms of selection can substitute for each other[10], there has been some agreement reached in social theory that the three types of rule-enforcing institution under discussion: critical and open discourse, competition, and (legitimate) authority, represent the **central social selectors**, that give specific rules and regulations selective advantage over continually arising alternatives. Success in the reproduction of social structures can usually be ascribed to the efficacy of these selection mechanisms or to a particular combination of them. The theoretical study of the prerequisites and consequences of these mechanisms stipulates the rather empty formalism of the "hard core" of the theory of action to the extent that social scientists are convinced of at least a reasonable amount of success in their theoretical work.

In contrast to this flourishing part of their research programme social scientists have made it much more difficult when it comes to the question of explaining how structural changes occur. The example of Marxist theory was obviously not strong enough to prevent a kind of equilibrium theory from becoming too dominant a paradigm in many theoretical areas. Classical national economy, just as social anthropology and sociological functionalism and systems theory can scarcely withstand the accusation of over-emphasizing the analysis of stable and balanced social relationships. Without doubt any social theorist accepted the empirical evidence of revolutions, structural collapses or unexpected failures of regulatory mechanisms as a falsification of this view. But within the conceptual framework of a consistently applied equilibrium theory it proved impossible to regard such events as anything other than the destruction of a formerly intergrating process, the selective power of which had up to that point succeeded in preventing, checking or blunting destabilizing action consequences. By logical consequence, any extensive structural change in such models of stability immediately had to be interpreted as a form of anomie (or pathology) without the slightest indication as to how noticable disintegration could be checked through new structural forms. As a result of the accepted assumptions, ordered change (i. e. the regaining of order and integration of social relations) could only be seen as a kind of "preordered structural differentiation" and the ensuing re-integration of different kinds of actions or subsystems, i. e. change was considered as a form of "structural growth", the success of which could be attributed mainly to endogenous forces of the changing systems (Schmid, 1983). Both conceptions of change (change as differentiation or as disintegration of structures and rules) originated in the logic of the theoretical models employed, which, for the sake of formal precision, were regularily involved in the reproduction of system-processes in the form of continuous functions. All those factors to which disruptive influences were attributed have consequently been treated as external "parameters", which one could choose to classify, either on an

ad-hoc basis or contrary to all experience, as being constant or as
striving towards some limiting value. Hans Albert vividly described the
lack of empirical content resulting from such procedures as "model-
platonism" (Albert, 1967, pp. 331-360). In view of this rather dismal
state of affairs, social science theory found itself faced with the
ultimately unacceptable alternative of either defending the non-empirical
character of its theoretical models, thus supporting a meta-theoretically
doubtful instrumentalism, or of resorting to stochastic models, thereby
interpreting the situational variability of social relations as a product
of fundamental and unavoidable imponderabilities in their reproductive
processes, and without giving a plausible or convincing empirical
explanation of this view.

Here the heuristics of a theory of structural selection is able to
propose some methodological remedies: First, the systematic study of
specific distributions of resources necessary for structural reproduction
can tell which options are open to a system of rules of conduct, should
the destabilizing effects of the corresponding social behaviour become
intolerable. In this context differentiations or the development of
subsystems turn out to be only one of several directions the actual
process of structural selection could take, as one can neither rely on the
fact that the corresponding organizational expertise will, in all cases,
be available, nor that the respective environments of a differentiating
system will welcome this kind of structural change (Schmid, 1982c, pp.
170-176, 226). This moderation of the classical theorem of differentiation
becomes additionally acceptable through the development of mathematical
models, that give plausibility to the idea that the transition from one
system of rules to another need not necessarily be a slow and gradual
process, but that, under certain conditions, it can occur in an abrupt and
discontinuous manner. Thus there can be irreversible bifurcations, step
functions or catastrophic changes (Woodcock and Davis, 1980; Zeeman, 1977;
Fararo, 1978; Renfrew, 1984, pp. 366-389) in the internal rule structure
of a system that do not inevitably entail an anomic collapse of all
regulatory mechanisms; rather, the way is then paved for the comparatively rapid (at least in terms of the timeperspective of a system's
reproductive process) emergence and establishment of a new rules
structure, itself capable of reproduction. The reliance on such models

allows one expressly to do justice to the undeniable historical experience of disruptive structural changes by correcting and modifying the traditional theory of social equilibrium with the help of hitherto unknown or unnoticed causal factors and generative processes. This modification showed exactly **why** equilibrium theory can be classified as being only approximately accurate[11] (which everbody already knew to be the case).

The heuristics of the theory of structural selection not only dispells the antinomies of the main ideas in the traditional theory of equilibrium, but also opens the way for new lines of research: First of all, it suggested investigations into the theoretically important and relevant differences of divergent selective processes and their inter-relations (Giesen and Lau, 1981; Luhmann, 1971b, pp. 370-371). Second, it suggested ways to apply these selective processes reflexively to themselves thus dealing with the problem of how they themselves were selected. Under what conditions, for example, would one choose authoritative as opposed to consensual decision-making or market organizations as opposed to authoritative rules and consensual procedures etc. Thus the "evolution of evolution" becomes a dominant theme in social scientific theory. Third, the comparative richness of selectionist heuristics allows for a more integrated perspective on different theoretical endavours: It becomes more and more evident that those rather materialistic theories, emphasizing material resources, technologies and organizational forms as the driving force of societal evolution, can be easily and fruitfully combined with predominantly "idealistic" approaches with their accentuation of normative orientations and value ideas in the selection of rules and institutions (Van Parijs, 1981; Corning, 1983, pp. 216-237; Corning, this volume; Eder, this volume).

V. CONCLUSION

I fear that, contrary to my original intentions, this article has become rather too technical. I feel however that one should not waste any opportunity to put forward either plausible arguments or useful examples which appropriately show how the central theory of collective action can be reconstructed whithin an selectionist framework. Even granting that the wheels of any theoretical transition turn only slowly, because academic life itself has to observe the rules of differential selection, I nevertheless do hope that selection theory may in due time gain some selective advantages as a unifying programme for social theory. After all, some of the most honoured social theorists such as Marx, Durkheim and Weber are already adherents.

NOTES

1. Both Spencer and Darwin accepted Newton's philosophy as a guide-line to their own theoretical endeavours, cf. Peel (1971, p. 14); Burrows (1966, p. 215); Ruse (1979, pp. 174-180); Ruse (1982, pp. 44-47).

2. As a matter of fact Spencer cannot be held responsible for all exaggerations of this approach, cf. Carneiro (1973a); Turner (1985).

3. Reproductability is based on some kind of autopoietic organization of systems (or organisms). But I shall not discuss this issue here. Cf. the idea of 'autopoiesis' and its meaning for evolution in Eder (this volume); Pieper (this volume), Wuketits (this volume), and Schmid (1987b).

4. It would digress too far to demonstrate in the context of the present paper that each social-theoretical discipline deals only with isolated and selected aspects of a more comprehensive action approach. For an outline of a possible unification of different theories which are usually considered to be seperated paradigms of action, cf. Glück and Schmid (1980).

5. These concepts stand for a time-honoured discussion in theoretical social science; cf. Marx (1965); Boudon (1979, 1980); Merton (1964, pp. 51-82); Moore (1967).

6. Cf. Giesen and Lau (1981), who at least implicitly show that my argument can be supported.

7. This has been the core problem of social theory since its very beginnings, cf. Platon (1977).

8. Ullmann-Margalit (1977, p. 134) argues that all kinds of distributive inequalities are fundamentally in need of external, normative stabilization and legitimation.

9. Parsons might rightfully be called the most important contemporary theorist of social integration by means of a common set of values, cf. Schmid (1982b, pp. 19-30).

10. Giesen (1980), argues in favour of a mutual substitution of these different mechanisms; Habermas is inclined to argue against.

11. I do not deny that there are equilibria (especially if we regard them
to be 'moving' equilibria), but to the extent that we understand them as
always depending on unstable relations between external parameters (cf.
Valjavec (1985)) we should be prepared to accept seeing these equilibria
dissolving rather quickly.

KLAUS EDER

LEARNING AND THE EVOLUTION OF SOCIAL SYSTEMS

AN EPIGENETIC PERSPECTIVE

I. EVOLUTION AND THE ROLE OF THE EPIGENETIC SYSTEM

The theory of epigenetic developments in evolution rests upon two assumptions. First, it refers to developmental processes that decouple biological from genetic evolution. Decoupling evolutionary processes from genetic evolution is even more important for social evolution. Second, it claims that the development of an organism plays a vital role in evolution. It takes into account the specific role individual development plays in evolution.

Thus epigenesis refers to definite evolutionary processes unintelligible within Darwinian theory (Ho and Saunders, 1982). This special characteristic of epigenetic processes restricts the field of random developments in evolution. The Darwinian processes of variation and selection are seen as of secondary relevance for evolution to take place. The logic of evolution is decoupled from Darwinian logic, which thus loses its pre-eminent role in explaining evolutionary sequences.

An epigenetic system that organizes individual development as cognitive learning processes (as does the epigenetic system underlying social evolution) changes evolutionary processes in several respects. It changes (1) the tempo of evolution (2) the internal structures that restrict the relevance of selection processes and (3) the conditions that favour learning processes and therefore the innovations that are necessary for social evolution.

The central characteristic of social evolution is that society is produced by such cognitive learning processes. Learning processes allow for the **self-production** (Touraine, 1973) of society. Of central importance

M. Schmid and F. M. Wuketits (eds.), Evolutionary Theory in Social Science, 101–125.
© *1987 by D. Reidel Publishing Company.*

to the processes of self-production is a special type of cognitive
learning, namely **moral learning** (Fairservis, 1975). Moral development
emerges in learning processes specific to the human species, and is
therefore considered to be the key variable in a theory of social
evolution (Eder, 1976, 1984; Habermas, 1981).

The Darwinian assumption that epigenesis has to be explained as a by-
product of the evolutionary mechanisms of adaptation through variation and
selection is incompatible with the idea of moral development through
social evolution. On the contrary, evolution on the social level is a by-
product of epigenetic processes of cognitive development. We claim that
the epigenetic system - being itself a by-product or end-product of
evolution on the biological level - to be the primary factor in social
evolution[1].

But this argument against Darwinian theory is still deficient. For it
can be argued - and this is the theoretical strategy of the behavioral
school of sociocultural evolution (Langton, 1979) - that the cognitive
learning capacity of individual human beings is subject to evolutionary
processes on the sociocultural level. Starting with such individualistic
assumptions Darwinian theory cannot be put into question. The theoretical
implication of this argument is that social evolution is the result of
variation and selection working directly upon human individuals.
Epigenesis is considered to be nothing but an obscurantist assumption in
the theory of biological and social evolution.

This argument against an epigenetic approach to social evolution can
only be refuted by showing that the ontogenetic acquisition of cognitive
capacities is necessarily a social process, that learning on the cultural
level is necessarily a collective learning process. Only when cognitive
capacities are shown to be socially constituted can the argument
concerning the central role of the epigenetic system in social evolution
be defended. The central thesis to be defended in the following is
therefore that the developing entity in social evolution is not the
individual, but culture. **Culture** is the result of the social interaction
of learning individuals, the result of a **collective** learning process.
Cultural development therefore reinforces the epigenetic processes that
make social evolution possible. The more cultural development advances the
less Darwinian theory, and with it each variant of individualistic

theories of social evolution (Schmid, 1982c), seems to be an adequate theoretical strategy.

My thesis about the role of evolutionary theory in the social sciences will be discussed in three steps. First, a classical attempt to explain evolutionary change in history through genuinely social factors (the Marxian argument) is reconstructed and then reformulated. Second, the idea of a social construction of the cognitive universe is discussed using the example of state-formation in historical tribal societies in Angola. In a third and last section, the theoretical consequences of the axiom of a social constitution of learning processes in social evolution and its effects upon a theory of social evolution are explored.

II. EPIGENESIS AND EVOLUTION IN SOCIOLOGICAL THEORIZING
2.1 TWO OLD ANSWERS TO THE PROBLEM OF SOCIETAL CHANGE

Why do societies change? Marx gives two different answers to this question, and these answers still generate competing views on social change. The first answer is that societies change because people relate to each other in an antagonistic manner; this is the theory of class conflict. The second answer says societies change because they are continually forced to adapt their own normative framework to changing environments; this is the theory of the structural strain between the productive forces and the social relations of production.

These answers lead to mutually exclusive evolutionary theories. The theory of class conflict rests on assumptions about collective action generating social change. This first answer is epigenetic insofar as Marx assumes the logic of collective action to be the logic intervening into social evolution. His second answer is Darwinian because in principle the productive forces select - under certain conditions - for specific social relations of production. This strain theory is compatible with Darwinian assumptions of social systems put under selective pressures in a given environment.

In recent reformulations of the Marxian theory the first answer has gained a new significance. In Habermas' reconstruction of Historical Materialism (1979) the evolution of society is conceptualized as the evolution of the **normative structures** of the system of society. The evolution of normative structures is to be seen as an epigenetic process based on an internal logic that is different from Darwinian logic. The unsolved problem in this reconstruction of Marx' first answer is how normative evolution comes about. Evolution is reduced to a developmental logic of normative structures derived from the logic of individual cognitive learning process. Its mechanisms remain unknown; they are to be looked for in "history".

But this separation of the logic of evolution from the mechanisms of evolution conceals the basic nature of this process: that evolution consists of the production of normative structures within processes of communication.

Normative structures change, not because of the cognitive capacities of some individuals, but because there is disagreement about normative questions between individuals. This conflict forces those tied to different views to communicate and then to learn how to coordinate their antagonistic orientations and convictions. The change of normative structures then is the result of necessarily **collective learning processes** (Miller, 1986).

Such a theory of the dynamic aspects of the process of normative evolution pushes an epigenetic interpretation of Marx' first answer one step further. The attempt will be made to reconstruct class conflicts as collective learning process and then to incorporate the idea of class conflict as the mechanism of social evolution into a general theory of social epigenesis.

2.2 SOME IMPLICATIONS OF THE THEORY OF CLASS CONFLICT

What is a class conflict? Class conflict implies antagonistic views about what course societal development should take, and is thus conflict over the cultural orientation of the development of society. Class conflict is the mechanism or, as Marx puts it, the motor of historical development. This definition allows for the preliminary distinction between two types of societies: into those with and those without class conflict.

Societies without class conflict are societies without history (Levi-Strauss, 1962). Research in social anthropology has given us a series of examples of social structures whose historicity has been destroyed. In such cases society can be regarded as a closed system of classification. A society has no history if it is totally classifiable in the terms of its own logic. These societies have "forgotten" past antagonisms; they exist in a closed cultural universe. An example of a traditional society without historicity is the caste society in classical India whose ideal structure was identified by Dumont (1967, 1970) as being based on the difference between pure and impure. There also are examples of modern societies claiming to be based on a stable social structure defined by egalitarian principles. Thus socialist societies claiming such a stability can be considered to live in such a closed cultural universe.

But the theoretical idea of a closed cultural universe and the related idea of a stable social order is based upon an illusion. It reproduces the illusionary image by which society describes itself. This self-description is characterized by the **suppression** of social antagonisms. The theoretical image of a stable society without class conflict reproduces an illusory social consensus, one where the **official** image of a society has succeeded in neutralizing opposing **unofficial** images of itself. In such societies, changes can only be induced from outside, be they demographic changes, changing material circumstances, or changes induced by the crisis-ridden logic of the societal system itself. In such societies, collective learning processes are necessarily blocked. These societies have to "wait" for the "objective laws of history". This critique of a certain theoretical image of society holds for primitive societies as well as for traditional and modern societies.

This critique implies that class conflict is the normal state of affairs. An adequate theoretical image of society has to start with the assumption that societies have a history. Society is not a classified and classifiable entity, but an action system within which opposing collective actors struggle for the control of the classificatory system. This antagonistic situation has been institutionalized historically in at least three different ways.

The first is through **ritual regulation.** Pre-state-societies institutionalize class conflict in the form of rituals. A collectively shared model of social organization is reaffirmed against opposing forces through ritual processes symbolically enacting the decomposition and recompostion of the social order. The ritual process is, as Turner (1969) puts it, an attempt to establish anti-structure in order to reestablish structure. It is a process negating the negation of structure. Ritual regulation in simple societies serves different functions at the same time; it helps to solve quarrels between families as well as ecological problems. It regulates interpersonal relations as well as environmental relations (Rappaport, 1979, pp. 27-42). But there is a much more fundamental aspect to ritual. It is also a mechanism for the reproduction of a social structure against the disorder brought about by group conflicts in the society. Rituals function to regulate conflicts between groups in a village and between villages. By regulating warfare and thereby reorganizing asymmetries between lineages, rituals guarantee social integration on the level of the societal system. Class conflict in pre-state-societies therefore can be said to be regulated by rituals.

The second way class conflict has historically been institutionalized is through **domination.** In pre-modern state-societies class conflict is controlled by legitimate political authority. In these "traditional" societies the continual display of the symbols of political authority, especially its hierarchical representation, allows for the regulation of class conflict. When this form of political authority is weakened class relations are then defined by the naked use of force to suppress the peasants, a quite unstable solution, as the history of classical empires shows. How class conflict is structured in traditional societies can be seen in Geertz' analysis of the 19th century Balinese kingdoms (1980b). The king, called Negara, was the guarantor of an order which bound together the antagonistic groups of society. He symbolized the holy order within which antagonistic groups struggled for control of power. This symbolic role was institutionalized in his official function: to secure the continuity of the ceremonies. That the ceremonial context is the field of class conflict in traditional society is also emphasized by Sahlins (1981, pp. 67-77). Geertz further argues that it is the symbolic universe that gives individual actions of a chief or a commoner their social weight: the position of an actor (be he a chief or a commoner) in the culture-as-constituted does determine the consequences of his individual actions. In the course of such culturally-constituted and socially-classified collective action the symbolic universe is transformed by fitting the intersubjective context to the given objective context. The ideal form of such a vertically constructed society has been found in the cultural system of hierarchy as exemplified in the Indian caste system (Dumont, 1967).

The third way class conflict has historically been institutionalized is through **permanent class conflict**. In modern societies neither ritual nor political authority can continue to serve as a social base for class conflict. A new structure is needed for the situation where class conflict has become a permanent one. The new institutional form is the democratic handling of conflicts, and this implies the self-production of society by collective learning processes. This is why modern society is the first society that can actually describe itself as a class society (Mousnier, 1974, pp. 13-46; Luhmann, 1985). Marx draws a radical conclusion: He conceives class conflict in modern society as **class struggle**. This concept of a struggle presupposes a non-normative concept of class conflict that can in fact be found most explicitly in the Darwinian concept of society (to which Marx sometimes seems to adhere). But there is also another model of class conflict contained in Marx' work. This model is based upon a normative concept of what constitutes class relations in modern society (Habermas, 1979). Marx gives some hints when he discusses the proletariat organizing a collective learning process in order to constitute itself as a class. As a learning collectivity the proletariat uses class conflict as the medium of "emancipatory" (in the strict sense of the word) learning processes.

The arguments concerning the theoretical categorization of class conflict are arguments for the idea that society produces itself via class conflict, that class conflict defines the dynamics of the process of the social self-organization of society. The category of class itself is an empty category; it refers to possible empirical referents of the process of social self-organization. In analysing concrete societies this category has to be filled with examples. Developmental systems whose dynamics are based upon class conflicts are socially structured. But nothing has been said concerning the structural properties of the situation called class conflict. To further my argument the structural properties of an epigenetic system regulating social evolution by the mechanism of class conflict have to be identified.

2.3 FROM THE DYNAMICS TO THE LOGIC OF EPIGENESIS

The idea of class conflict developed so far gives some preliminary hints
concerning the operation of an epigenetic system within the process of
social evolution. The next step consists of abstracting from the idea of
class conflict. The epigenetic system is to be constructed on the level of
a general theory of social action. It will be shown that the difference
between Darwinian and epigenetic assumptions has to do with two
incompatible theories of action. The first theory conceives social action
to be guided by the calculation of anticipated profits. The second
conceives social action to be constituted by communication. The first
theory is of a psychological, the second of a sociological nature, a
difference that Campbell (1976) interprets as a conflict between
psychology (science) and moral tradition (religion). Psychology is indeed
inadequate to treat moral traditions scientifically; for here we have to
deal with a genuinely sociological phenomenon. That leaves us with the
question of science of morals, **i.e.** sociology.

The **"action-as-profit" theory** (Harris, 1979) assumes that changes in
social structure are dependent upon solutions of the biological,
psychological and ecological problems experienced by all human beings and
all human cultures. The logic of action is utility: People prefer those
situations which work to their advantage. Thus in a situation
characterized at the same time by ecological barriers and growing
overpopulation where war is a normal consequence, less powerful groups
might well elect a permanently subordinate status. The benefits of such a
status can be said to exceed the costs of trying to maintain independence
or retreating into ecologically less favorable environments. The logic is
simple: Benefits and costs are calculated in terms of natural needs like
hunger, survival etc. Culture is nothing but the collective effect of
aggregated individual strategic actions. This behavioural theory of action
does not distinguish between epigenesis and evolution. Individual action
is coordinated by selective pressures upon individual actions; the logic
of coordination is reduced to the logic of selective pressures. The only
problem left is to define the selective structures as such. In this theory
epigenesis is without importance for evolutionary changes. Epigenetic
assumptions are nothing but a form of obscurantism (Harris, 1979).

The **"action-as-communication" theory** (Leach, 1976) assumes that social change is dependent upon a shared symbolic universe which allows for communication. Symbols order the world for the people concerned. Shared interpretations of symbols generate a world in which men can communicate with each other. Culture is a shared symbolic universe (Geertz, 1973) that allows for the coordination of social action. This theory assumes that without the collective construction of a symbolic universe there is no object upon which selective pressure can be exerted to produce **social** evolution. The theory therefore separates the development of a shared symbolic universe from the evolutionary pressures upon a symbolic universe. Epigenesis then becomes of utmost importance in sociocultural evolution.

A behavioural theory of action has no need to know how culture is organized. A behavioural analysis analyses nothing but the actual performance of the members of a culture. But this observable performance could be a bad performance of a cultural script. To go beyond a behavioural theory therefore the plan of cultural has to be known. Leach gives an illuminating analogy: To know the score of a symphony it is not sufficient to observe the performance of that symphony by an orchestra; you also have to know the rules that are obeyed by the musicians. The same is valid for culture as a whole: To know a culture requires knowing the rules underlying the actions of its members (Leach, 1976). This implies looking into the organizing structures of culture in order know the epigenetic logic of social evolution.

The structuralist movement in cultural anthropology has identified some general structures of cultural systems. Fundamental to structuring cultural systems is **binary classification** (Levi-Strauss, 1962), a principle that is characteristic of all cultures. A second fundamental rule is **ranking** (Schwartz, 1981). This relational structure underlies the imaginative ordering of nature as well as the normative ordering of society in diverse cultures. Whereas binary classification is a cognitive mechanism constituent of culture, ranking refers to a cognitive mechanism that allows for vertical social classification. Thus ranking articulates more specific differences between forms of a social order.

Three general types of ranking underlying social configurations can be distinguished: analogical, hierarchical and functional. Each has different structural effects upon the social construction of reality. **Analogical** ranking is typical for natural communities; ranking is here bound to natural roles. **Hierarchical** ranking is based upon more complex images of nature, those that distinguish between the natural and the supernatural through a superordination/subordination relationship. This type of ranking sees the social order as the extension of a natural order. Social ranking is based upon one's place in a hierarchical and therefore holy order; the idea of caste is the clearest case in this respect. **Functional** ranking sees the social world from "individualist" premises. The image of an ordered "super-nature" is replaced by the image of cooperating individuals. Social ranking is based upon individual rights or individual success. The idea of an egalitarian society of free individuals is the ideal type of a society organized along these lines.

These different cultural logics are not restricted to the function of organizing cultural representations of society: The symbolic structures that make up the different cultural worlds are not only created and changed by rules of communication, ideally by **rules of argumentation** (Miller, 1986). They do something much more fundamental: They make communication in society possible by regulating a specific property of communication, the possibility of saying "no". Communication necessarily produces conflicts because people can say "no" to a communication. This specific property of social communication reveals the function of a shared culture: to define a collectively shared symbolic world that restricts the possibilities of saying "no" which is the precondition for entering into a process of resolving conflicts. Through resolution the shared world can be changed and expanded and can serve a new as a reference world for future conflicts.

In pre-state-societies the logic underlying the resolution of disputes reflects the age, personal prestige and status of the parties involved (Gluckmann, 1977). Natural differences make up the social structure of this type of conflict resolution. A second form of handling disputes can be found in hierarchically organized societies, where the tacit acceptance of a hierarchy is the cultural presupposition common to those engaged in a dispute. The word of the king has more weight than the word of the peasant. In such societies, the appeal to authority is thus the traditional solution to the problem of conflict resolution. A third way of resolving conflicts is based upon egalitarian norms. They provide the means for a form of conflict resolution that allows for the neutralization of inequal status. This is the modern solution to social conflicts.

These three forms of conflict resolution constitute three distinct forms of social order. The first can be interpreted as producing a concrete interactive morality underlying a social order, the second an authoritarian one, the third an egalitarian one. From an evolutionary point of view these types of morality can be seen as stages of the evolution of a social order. The stage-like change of morality can be said to be the outcome of an epigenetic process. Social evolution can be said to be bound to a **moral epigenesis** that passes through a pre-authoritarian, an authoritarian and a post-authoritarian stage of conflict resolution.

This stage theory should not be confused with those stage theories that derive the properties of stages from institutional properties (Fried, 1967, 1975; Service, 1975; Cohen and Service, 1978). Institutional forms are the time- and space-specific realizations of a moral order. They are the result of evolutionary pressures upon moral orders and are therefore not indicators of moral stages. Whether or not these epigenetic assumptions are necessarily bound to the assumption of a developmental logic (Schmid, 1982a) is a problem to be discussed later. That there are normative implications in such a theoretical approach cannot be denied.

III. EPIGENETIC DEVELOPMENTS AND SOCIAL EVOLUTION
3.1 Stage Models in Evolutionary Theory

How are epigenetic developments related to evolutionary processes? Epigenetic developments are not independent of those evolutionary processes that are defined by transformations over time and variations in space. Epigenetic processes cannot be separated empirically from Darwinian evolutionary processes.

The Darwinian conception of evolutionary processes does not imply any assumptions about stages. The specific evolution of a concrete social system in space and time depends upon the internal properties of that specific system and its place within an larger natural and social environment. In this sense it is possible to describe the specific evolution of a tribal system into a more complex system, that is into a state-society. The logic of the epigenetic system is reduced to the logic of the systemic functioning of society. Friedman and Rowlands (1982) have made an impressive attempt to do this. Using the example of the evolutionary change of a specific type of tribal system they start with analyzing the social relations of production and exchange and try to show how internal trends toward increasing complexity over time can either be neutralized or differentially selected for by a given structur of the environment.

This concept of an epigenetic system forces Friedman and Rowlands to introduce the difference between specific evolution (which is epigenetic and local!) and general evolution which refers to the development of the spatial system and the structural time period within which epigenesis can take place. The logic of general evolution is the dominant logic; it is the logic of a system reproducing itself in a larger environment and thus being independent of the logic of the epigenetic system. This evolutionary logic determines the degree of change necessary given the initial structures of the local system in question.

But beyond the degree of transformation is the problem of the very character of this transformation. Do the changes in structure allow for collective learning process or not? Assuming that human collective action "normally" implies learning processes on the part of those engaged in it, we can discriminate between evolutionary changes that bring about learning and those that do not. A theory of evolution that overlooks the possibility of learning or non-learning as an outcome is forced to subordinate epigenetic processes to the factors of space and time that select for transformation. The possibility that a society will not learn can never be ruled out. Change in a society that does not learn, must be accounted for by selective pressures exerted upon it. Thus an anti-epigenitic evolutionary theory is limited to accounting for very specific cases: societies that do not learn.

There is a real problem inherent in an evolutionary theory that reduces the epigenetic system of systemic properties of local systems defined in space and time. It underrates the role of learning processes for the evolution of society. The problem how to relate epigenetic process (i.e. learning processes) to evolutionary processes (in the strict sense, i.e. of the Darwinian type) in a more productive manner will be

treated in the following using an example that has become the object of
numerous evolutionary explanations: the origin and evolution of the state
in the history of mankind[2].

3.2 Evolutionary Theory and the Problem of State Formation

The process of state-formation can be described as the evolutionary
transformation of tribal systems into state-societies. A "state" implies a
normative framework that reorganizes on the most fundamental level the
kin-society, the form of social organization typical for pre-state-
societies. This process has been the subject of diverse attempts to
construct an evolutionary theory of social change (Fried, 1967; Service,
1975; Eder, 1976; Claessen and Skalinik, 1978a, 1978b).

Currently, the dominant theory of the formation of the state is based
on the cultural evolutionist model. This theory starts with the assumption
that pre-state-societies are characterized by an "egalitarian" form of
social integration based on kinship relations. The kinship structures
underlying these relations regulate hereditary succession, access to land
and water, collective cooperation in the more important economic
activities and the distribution of goods between and within different
descent groups. Intensification of production and demographic growth lead
to shifts in social organization and to the crystallization of the role of
"big men" who represent more complex forms of political power.

The appearance of big men allows for the institutionalization of
functionally specific decision-making procedures, procedures which are
much more flexible than those ordained by any kind of ritual regulation.
On the basis of this political power the big men can also theoretically
accumulate economic power. In order to uphold the old equilibrum (based on
reciprocity between descent groups) big men are supposed to organize
redistributive processes. They in fact have to be generous, have to give
away all their economic power in order to uphold their political power.
Should the population grow further and cause geographic and/or social
circonscription to tighten dissociation of political from economic power
will lose effect. Under such circumstances the generous redistributors can
reinforce their political power by transforming voluntary contributions to
the stock for redistribution into some kind of taxes. On this new economic

base big men are able to pay a clientele. The big men are transformed into
warrior chiefs (Sahlins, 1962). While these warrior chiefs are still bound
into a hierarchically organized kinship structure, they can now mobilize
their fellows for warfare and raids. Theirs is the power to substitute the
small patrilineal groups with multi-village military alliances. The role
of the warrior chief is the nucleus for the role of a king who
redistributes only in part those goods he has received by coercion. When
the role of the king is institutionalized primitive society is transformed
into a state-society.

In this model the concept of culture is based upon a naive theory of
social action. Henderson (1972) has succinctly stated the limitations of
this old evolutionary paradigm. He sees it as a theoretical simplification
which allows for a

> systematic concentration upon factors that may be
> called "external" to individuals: (a) social factors, or the
> constraints imposed by a few major types of socially structured
> situations (economic, political, ritual, etc.), and (b) ecological
> factors (the relationship between technology and environment). By
> setting all human social behaviour within a comparable structural
> framework, and assuming that each actor acts simply to maximize
> his own wealth or power and orders his learning processes towards
> this end, the scholar may readily direct attention to the social
> and ecological constraints that either produce equilibra within
> and between groups or else tend to change their structures
> (Henderson, 1972, pp. 3-4).

The theory of action underlying cultural evolutionism is the "action-as-
profit" theory. From these individualistic premises the social processes
leading to state-formation cannot however be grasped. The normative
structures within which strategic action has to take place are of no
theoretical relevance. Thus the main aspect of the transition from pre-
state-structures to state-structures, i.e. the transformation of the
structure of the social conditions of strategic action, remains hidden.

In the following chapter the evolution from pre-state-societies to
state-societies will be analyzed in more detail in an attempt to show how
epigenetic developments and evolutionary pressures interact. Of special
importance will be the attempt to prove the relative independence of the
epigenetics system (i.e. its development) from adaptive pressures (i.e.
evolutionary processes). Material from an ethnohistorical study of state-
formation (Miller, 1976) will be used to show how individual actions and
the normative modes by which they are coordinated are involved in the

transformation of social systems as basic institutions of society change. This will serve as a starting point for an alternative theory of state-formation as well as an argument for a radical epigenetic approach to social evolution.

IV. AN EPIGENETIC THEORY OF THE FORMATION OF THE STATE
4.1. The Historical Formation of Early Mbundu States

The empirical basis of the ethnohistorical account of state-formation in Angola from the 16th to the 18th centuries given by Miller (1976) is oral traditions. This allows Miller to speak not only about events, but also about the ideas related to these events. Oral history provides better empirical data for the construction of a theory of state-formation than are usually used in this field. The historical perspective allows Miller to reconstruct state-formation as a continual give and take of different pre-state political institution in the evolution toward a state-like structure. It shows how different groups use new political ideas either from their own social context or from an alien context on order to construct more cohesive political institutions (hunting groups, ritual group **etc.**). The decisive historical step is their transformation into states (**e.g.** kingdoms). This process succeeds to a certain extent. It is followed by break-aways from kingdoms which are then organized on the basis of differing local conditions. State structures are then modified again.

Such an "internalist" explanation of primary and secondary state-formation is directed against all "externalist" theories of state-formation, in this specific case against the theory of the Hamitic origins of African states and their "daughter" states explaining state-formation by migration and conquest by people with higher civilizations. These theories have survived in the so-called "Sudanic state hypothesis" (Miller, 1976, pp. 4-11).

Attempting to reconstruct the material of Miller for an evolutionary theory of the social origins of the state the following three points are to be stressed:

(1) Political institutions which cut across the lineage base of society were myriad in Mbundu pre-state-societies. Their functioning

depended upon the functioning of the basic units: the descent groups which regulated the material life, the land rights, the work process, and the distribution of goods within the lineage.

(2) Authority was conceived in a specific way by the Mbundu, resting on the ability to invoke supernatural sanctions. It was not inherent in human beings, but resided in authority emblems associated with titles. Authority was an abstraction, independent of its living incumbents. Thus authority was dissociated from concrete social relations; it has already become an ontological idea. This is to be taken as a criterion for an authoritarian morality.

(3) The institutional steps toward statehood in Mbundu societies can be seen as a process of socially constructing a generalized authority role. First authority is restricted to authority over persons other than kinsmen; it operates outside the kin-society. Then the king himself is made an outsider; he is credited with supernatural (magical) means and is given a certain secular institutional backing (slaves). At last the king becomes the impartial arbiter between competing lineage groups, basing his power on a legal right. The social construction of generalized authority roles follows a developmental pattern starting with a chief still being dependent upon his descent group, then moving through a   theocratic ruler up to a legally defined king.

The first point decribes lineage structure as hampering the development of political institutions into more enduring institutions. The descent group is seen here as the great conservative factor in political evolution. External factors such as the existence of salt pans or ports of trade contribute to the crystallization of more enduring political institutions than those typical for the lineage based society. But these "external" factors were never strong enough to enable the new political institutions to transform the descent structure into a structure better suited to the function of political domination.

For the Mbundu, ngundu, the descent group, was the fundamental mechanism of social integration. Kibinda, the hunting society, created links between the ngundu, and performed serveral functions essential to their welfare. On the other hand the kibinada cut across the ngundu and was in fact the nucleus for the beginning state-formation. The structural problem throughout the history of early state-formation is the relationship between such cross-cutting institutions and the lineage structure of society.

This structural problem is linked to the second point made above. Authority in Mbundu society was dependent upon authority titles or emblems: to have authority was to have control over an authority emblem. Historically, once a new symbol of authority had spread among the Mbundu lineages, individual holders were able to expand their personal spheres of influence, thereby appropriating authority over persons not related to them by kinship. In Mbundu society the decisive developmental step was the structural shift from lunga titles of authority to mavunga titles of authority. Lunga titles were hereditary titles, awarded to the lineages by lunga kings. Lunga kings were thus under the control of the lineage groups. These titles did no more than reinforce the links between descent groups regardless of the physical distance. The lunga concept of an authority role was still grounded in concrete interactions.

A different concept of authority was introduced with mavunga titles which were awarded to persons obliged to perform specialized duties in support of the king and his court. Mavunga titles defined for the first time among the Mbundu a social position lying outside the control of descent groups. These titles thereby created tensions between the lineages and the holders of mavunga titles.

Mavunga structure had its counterpart in the kinguri structure. The kinguri title derived from a specific lunga title. Kinguri groups tried to eliminate lineages as the organizational backbone of the social structure, and to replace the laws of descent with the laws of the kinguri. These laws were intended to hinder the establishment of descent relations and in fact forbade childbearing. Children would enter the band only through adoption or enslavement and would owe allegiance only to the kinguri. The kinguri groups moreover demanded total obedience, seeking thereby to abolish other competing titles of authority. This would lead to a centralization of authority in the kinguri. Historically, the creation of total power as the basis of kinguri state-building failed: The kinguri solution was unacceptable to other chiefs within the reach of the kinguri groups. These in fact broke with the kinguri and maintained their authority upon other lunga titles.

Both the mavunga and the kinguri images of authority were based more upon master-servant (or patron-client) relations than upon social relations through descent. They were attempts to create a consistent, abstract conception of authority apart from the ngundu descent relations. But these concepts lacked the supernatural legitimacy. The realm of the supernatural was still in the hands of the diviners, not the authority holders. With the rise of the Imbangala kings, who replaced the kinguri type of state system, this changed. Historically, the Imbangala kings constructed a universal moral difference between themselves and their subjects. The Imbangala kings represented themselves to be non-human in contrast to other people who were merely human. These kings ritually ate human flesh while forbidding it to non-king individuals. This cannibalism drew on analogies which the Mbundu saw between cannibals and carnivores. This ritual became the mechanisms of a hierarchical ordering between those with authority and those without it.

The early Mbundu states that based their political authority primarily upon lunga titles must be described as **chiefdoms**, because they were dependent upon lineage structures. The kinguri states and the Imbangala kilombo crossed this chiefdom level. They were theocratic states based upon elaborate rituals in which the symbols of basic social differences (esp. cannibalism or the separation of women from the world of political authority) were the elements of a new hierarchical social order.

The third point refers to the institutionalization of authority. The kinguri example shows that destroying the lineage structure does not suffice. The kilombo is an example of an institutional device that attempted to realize a consistent ideal of unquestioned authority. The kilombo was at the outset a circumcision camp, one of the many institutions cutting across lineage groups found in the societies of this region. This institution was used by Imbangala kings in an effort to establish a non-lineage social structure, thereby institutionalizing their hegemony over the Mbundu people. The initiation of new members into the kilombo through rituals not connected to kinship allowed a radical ideological break with the kin-world. With only one restriction, that these males not be circumcised, a requirement that qualified all those who had not yet undergone circumcision in their own lineages, the Imbangala kings were able to attract a considerable manpower. The uncircumcised

young men were not yet fully-socialized Mbundu. They were also young
enough to become easily indoctrinated into the culture of the kilombo.
Using the material and ideological ressources of the kilombo gave the
Imbangala king double control over his subjects, over male society and
over the relations with the supernatural. He was not just a chief, but a
theocratic ruler, fulfilling as well the function of the diviner that
traditionally had been preserved by designated diviners.

But the kilombo ultimately failed to build up a new infrastructure for
society. For to reproduce itself it had to rely mainly upon men coming
from Mbundu villages. These man reintroduced the old Mbundu ideas and
structures into the kilombo. The primacy of non-kin and non-human liaisons
was softened. The Imbangala ideology of non-humanness and the rigorous
conditions of life in the kilombo which differed dramatically from the
everyday life of the people continually worked against what the kilombo
was trying to establish. The new cultural system was only partly
institutionalized; for only the ruling class, not the dominated people,
accepted the authoritarian morality.

After the dissolution of the kilombo the Imbangala kings took another
step in political evolution with the help of the Portuguese. Slave trade
and legal backing by the Portuguese now became the basis of the Imbangala
kings. This "external" factor gave rise to a new relationship between the
Mbundu lineages and Imbangala kings which was reinforced by a consonance
between the ideology associated with the titles of the Imbangala kings and
the given Mbundu political system: The kings based their authority upon
the lunga titles the Mbundu were accustomed to. The lineages could banish
the kilombo, but not the cooperation between the Imbangala kings and the
Portuguese. After the end of the slave trade in the 1850s the balance of
power shifted again, back from the kings and in favour of the lineages.
This was the end of the traditional state for the Mbundu.

4.2 A Social-Evolutionary Theory of State-Formation

The example of the Mbundu shows that the political "moves" of chiefs,
leaders **etc.** are bound to a normative framework that puts restrictions on
strategic actions. Only within a normative context is strategic (or
utilitarian) action possible. In order to extend the range of possible

strategic actions the normative framework must be changed. Seen from this perspective the optimum of possible strategic actions is reached when the normative context itself is built upon the rules of strategic action. This is the state of nature described by Hobbes; it is society free of norms.

This type of evolution is always possible, but it never has a stable outcome. Stable solutions to the problem of a normative order of society must be built upon social structures that allow for a moral resolution of conflicts. For the Mbundu, such a social structure was only partially generated. But such a solution hints to the general conditions favouring evolutionary change toward state-societies. In the transition from simple to more complex societies the evolutionary prerequisite is the substitution of the logic of concrete reciprocity by the logic of political domination. Concrete reciprocity relies upon natural differences for its own legitimacy. Political domination relies upon hierarchical differences. This transition from pre-state-societies to state-societies succeeds because it transforms natural differences into hierarchical differences. Both differences are morally justifiable. Whether a social system is grounded upon natural or hierarchical differences depends upon what kind of morality socializes it.

Such a perspective implies a reversal of the perspective of classical cultural evolutionism. A big man is not transformed into a warrior chief because social changes give him the strategic chance to accumulate power. A big man becomes a warrior chief because people redefine a social situation in such a way that his social role can be played differently: as that of a ruler. The social situation is defined no longer by the logic of concrete reciprocity, but by the logic of political domination. This redefinition implies a complementary redefinition of the role of the people: They become the subjects of a ruler. But why do people change their definition of what they regard as the right state of affairs? What makes people change their mind in such a way that culture can be completely reversed?

The conjectural causes that set into motion or inhibit this process are to be described as factors external to the social actions of the people concerned. External circumstances that select for evolutionary change are a necessary condition of such a change. Pressure from the external material environment is necessary in order to set structural redefinitions into motion. Such a situation is identical with what Marx understood as the crisis of a mode of production. According to this theory, the formation of the state presupposes a crisis of the "neolithic" mode of production. Such a crisis can be described as the inability of the kinship structures to solve the problem of distributing the goods produced in a society. Insofar as kinship ties can no longer serve as an institutional frame for systemic reproduction, changes in the system become necessary.

Beyond these conjectural causes that function as selective mechanisms in an evolutionary process, some internal conditions on the cultural level also need to be fulfilled. They can serve as starting points for attempts to redefine the normative framework within which a society can be reproduced, especially solve its distributional problems.

Such conditions can be shown to function in the legal and religious field in the transition to state-societies. In egalitarian systems the legal function is integrated into the ritual complex (Koch, 1974). Chiefs have the right to act as arbiters. They even may have the right to sanction. But the only difference between the vengeance of the chief and that of any other person is that the vengeance of the chief is more cruel. Neither right is therefore specific to the chief. In stratified societies the legal function becomes more specific. In the kingdoms of the Shanti, Barotse, etc. we find judicial courts which are the property of chiefs or kings (Gluckman, 1977). But the chief or king is himself restricted to a symbolic function. While domination is accepted on a religious plane, the king is powerless on the legal plane; the legal function (sanction) is taken over by representatives of the king. This is an ambivalent solution that makes concessions to a concrete social morality, but also contains elements of an authoritarian morality of law and order. Only after the king is defined as the final judge can authority and domination be institutionalized (Eder, 1976).

This development of the legal culture has repercussions in the development of the religious belief system. The gods become masters of destiny; they are no longer the objective causes of the fate of the people. The gods become guardians of justice; they are no longer merely the guardians of the objective order of the world. New gods of law supersede the old chthonic gods of the archaic world. These new gods represent a new cultural logic, that of a hierarchical moral order. The cultural changes produce a new developmental system which is the starting point for the further social evolution of the social system of society. The legal and religious definitions of hierarchy imply a reorganization of the internal environment, of the epigenetic system of culture.

Such cultural change can be interpreted as a learning process. Learning processes are on the one hand processes by which the objective world is cognitively appropriated. As such they allow for the cognitive assimilation of external pressures upon action. But the result of learning processes such as new technical inventions also pose new problems for the reproduction of social systems. They force the moral infrastructure of society to readjust. Such a cognitive interpretation of cultural development as a learning process has already been proposed by Fairservis (1975) and others.

The "external systemic environment" contains the conditions that favour possible evolutionary changes. The "internal cultural environment" contains models for a social order. Each of these environments is behind one of two competing evolutionary theories, one an objectivist theory that explains the social forms of communication through selective pressures at work in society, and the other a subjectivist theory that explains the products of these social forms through an innate progressivism in humanity. The controversy between the two is wrong and misleading. In the first case, culture appears as a superstructure, representing reality more or less. In the second case culture determines social evolution. Both approaches are inadequate, however, because they each fail to take into account the interactive relation between culture and social form. Culture is socially produced, and in this process of social production society produces itself. In this process of self-production external factors may intervene and produce a social system determined by space and time.

The topological image of base and superstructure is also misleading. There is neither a special logic to the system of social relations (of production or distribution etc.) nor one to the system of cultural symbols, be they of religious or secular nature. But the Marxian tradition provides an alternative image to this misleading one: the image of class conflict as the motor of history, as the mechanism of social evolution. This image covers the idea expounded above exactly: Class conflict is at the same time a conflict concerning the cultural orientation of societal development and a social relationship that relates classes of people with each other. Class conflict is a mechanism that changes (if there is change at all) cultural orientations as well as the social context within which classes relate to each other.

This brings us nearer to an answer to the question of why people change their moral outlook. How can we explain the learning processes that react to external influences and at the same time produce a different cultural universe? This question can be answered with the concept of collective learning processes. For when moral questions are at stake, people have to communicate with each other, they have to enter into social relations, and this creates a reflexive mechanism. In trying to learn, people create a social universe within to organize their learning processes. And they create social contexts that cut across the established forms of social relationships. Such leveling social forms (e.g. religious rituals or hunting societies in pre-state-societies) are therefore marginal with respect to the dominant social environments (e.g. kinship institutions in pre-state societies). As they are used in handling moral disputes social forms become constitutive conditions for their resolution. In some cases this handling breaks apart the social form in which it takes place. These are the historical situations where new forms of social relations are invented.

Thus the class of people communicating in ritual hunting societies and the class of people communicating in kin communities represent antagonistic cultural orientations and contexts of communication. But as soon as these contexts enter into relation with each other a dynamic is set into motion that changes the cultural legitimacy of social forms of communication that have so far been uncontested. Learning has no need for externally induced dynamics: Already, before selective pressure comes in, the social world is in the state of epigenetic evolution.

V. CONCLUSION

The concept of epigenesis has a well-defined status in the theory of biological evolution. It refers to organic developmental processes that decouple biological evolution from genetic evolution. These processes are supposed to exert an autonomous role in evolution. Epigenesis on the social level does the same: It decouples social evolution from organic evolution. This is why this concept has been used to construct a theory of evolution that takes into account the properties of social evolutionary processes (themselves evolutionary emergent).

It has also been shown that the individualistic assumptions underlying the biological conception of epigenesis are a block to an adequate understanding of social evolution. Therefore the idea of an epigenetic developmental system governing evolution has been expanded into one that could better be called an **epiorganic system** governing social evolution. Social evolution is characterized by the fact that it has become possible to disconnect cultural development from individual learning. There is no requirement that all individuals learn. It is sufficient that they communicate. Through the process of communication some learn and thereby redefine the collective knowledge and consciousness of society. In this sense the specific characteristic of social evolution is that **society learns.** This often contested Durkheimian idea is to be defended against all forms of individualistic reductionism.

There are only two assumptions behind the theoretical idea of a sociocultural epigenesis: that disputes over moral questions are normal and that the resolution of disputes implies communication. As such this theoretical ideal can be formulated in a parsimonious manner similar to the old neo-Darwinian theory. Progressivist assumptions are not needed. Therefore the classic objections to applying epigenetic approaches in theories of social evolution are unwarranted.

Our theory of an epigenetic evolution thus changes the function that the idea of variation and selective retention can have on the sociocultural level. It also modifies the function that the idea of a developmental logic (leading to an end state) can have on social evolution. For these ideas are themselves descriptions of social evolution that fulfill different functions. As the idea of sociocultural selection, Darwinism is an attempt to relate the status of a society in the international system of societies to its competitive strength. As the idea of a developmental logic in history, progressivism is an attempt to mark off the distance between the past and the present in terms of primitive versus modern and thus to evaluate the status one society has vis-a-vis another.

As soon as these theoretical positions are disputed the social context changes. Once these positions are seen as self-descriptions of society, the vantage point for observing the system also changes. The position of the Darwinist or the Progressivist has become obsolete. For we have learned that these are nothing but possible ways of looking at society. They have become antagonistic ways of society looking at itself. Society - after the disillusioning experiences with its classical theoretical self-descriptions - has to reorganize its description of itself. When we try to observe this new antagonistic reflection we are left with the sole idea that what we observe is nothing but the collective learning process in which we as the observers also take part. What we observe is nothing but the "autopoiesis" of society in a collective learning process in which we take part.

NOTES

1. The concept of "evolution" is used in its strict meaning: The term evolution refers to stochastic process generated by the mechanism of mutation and selection. The term epigenesis refers to developmental processes, especially to the processes of cognitive and moral development in social evolution. Therefore speaking of moral or cognitive evolution is somewhat misleanding.

2. The literature on state formation is vastly expanding. For some of the more important recent literature dealing with this special topic see Wright (1977), Saxe (1977), Claessen and Skalnik (1978a, 1978b), Claessen (1978), Skalnik (1978), Cohen (1978), Bloch (1982).

PETER A. CORNING

EVOLUTION AND POLITICAL CONTROL
A SYNOPSIS OF A GENERAL THEORY OF POLITICS

I. INTRODUCTION

In the past several years, there has been a broad-based renewal of
interest in the explanation of social change. One manifestation of this
trend has been a return to the "roots" of human society, in an effort to
gain greater understanding of the causal dynamics underlying socio-
cultural evolution. Though much new data and many insights and hypotheses
have been added to our store of knowledge about the evolutionary process,
no general theoretical synthesis has so far been attempted. Some theorists
maintain that no such synthesis is possible - that the subject-matter is
inherently too complex and refractory, or that a synthesis at this point
is premature. I obviously disagree. I believe the time is ripe.

Let the reader be forewarned at the outset, however, that what follows
is, of necessity, an abbreviated summary of a much larger monograph
(Corning, 1983). The reason why the theory that I am proposing cannot be
treated in detail here is because it involves, not one, but three
interrelated theories. A major component is a general, empirically-
grounded (and, I believe falsifiable) theory of politics. But this theory
is, in turn, an integral part of a theory of cultural evolution, which is
in turn a special case of a theory that attempts to account for the
"progressive" evolution of biological organization generally.

These interconnections are not arbitrary, and there is nothing
gratuitous about this "super-theory", for I am addressing a fundamental
aspect of a single - albeit multi-faceted - historical process. Mindful of
Albert Einstein's observation that "a theory is the more impressive the
greater is the simplicity of its premises, the more different are the
kinds of things it relates and the more extended its range of
applicability," it is my hope that it may be possible to "reduce" certain

127

M. Schmid and F. M. Wuketits (eds.), Evolutionary Theory in Social Science, 127–170.
© 1987 by D. Reidel Publishing Company.

aspects of the evolutionary process - both in nature and in human societies - to a unifying theoretical framework. Indeed, I believe that the key elements of the theory can be stated very succinctly (as I have attempted to do in the abstract above).

Explicating this theory in detail, however, is another matter. In order to keep the discussion within bounds, I will confine myself here to a brief summary of six key aspects of the full monograph: (1) a general statement of the problemset, or theoretical "puzzle" to which the theory is addressed; (2) a view of evolutionary causation that reverses our conventional understanding of cause and effect relationships; (3) a discussion of principle of functional synergism, which I posit as the underlying cause of the directional evolution toward more complex forms of organization; (4) the case for the cybernetic model (properly understood) as a general model of the decision-making, communications and control aspects of complex goal-oriented systems; (5) an overview of how these elements fit together into a theory of politics; and (6) a summary of how this theory relates to two of the major domains of political science - political development and international relations.

II. THE THEORETICAL PROBLEM

Regarding the first aspect - the problem-set or theoretical puzzle - it is important to keep in mind that what I am proposing is a theory only about a particular facet of evolution: biological, cultural and political. I am addressing the problem of accounting for the directional, or "progressive" evolution of complex forms of organization; it is a theory about the evolution of "systems". It is not a theory about "everything", either in nature or in human societies. Thus many important theoretical problems are not addressed directly. I do not attempt to explain why dinosaurs evolved and then became extinct, why Mayan civilization arose and collapsed, or why the United States came into being and evolved as it did. The precise explanations for such events should fit within my framework, just as the precise causes of evolutionary change generally should fit within Darwin's natural selection theory. However, the general explanatory principle must always be combined with various context-specific factors to provide a complete explanation. An analogy is to the relationship between Newton's theory of universal gravitation and the problem of explaining how a particular leaf from a particular tree happens to flutter to earth at a particular time and place. Gravity explains the general direction in which the leaf will travel, but many more factors - physical, biological and "historical" - also contribute to determining the precise "event".

In effect, my focus is upon Herbert Spencer's 19th Century vision of evolution as a unified process of progressive "complexification" (a vision shared in varying degrees by many contemporary theorists). In a nutshell, I attempt what eluded Spencer. That is, I develop a causal explanation of "progressive" evolution (and its antipode, "regressive" evolution) that is compatible with Darwin's theory of natural selection. In Spencer's definition, evolution involves "a change from a indefinite incoherent homogeneity to a definite, coherent heterogeneity through continuous differentiations and integrations" (1862, p. 216). "From the earliest traceable cosmical changes down to the latest results of civilization, Spencer wrote in **Progress, Its Law and Cause** (1892 (1857), p. 10), "we shall find that transformation of the homogeneous into the heterogeneous is that in which progress essentially consists." Accordingly, Spencer posited a "Universal Law of Evolution" based on what he portrayed (at least in his early writings) as a deterministic, self-propelled trend. For a more recent understanding of "progress" see Sahlins (1960).

Notice that Spencer equated evolution itself with a specific directional trend (as do a number of contemporary anthropologists and sociologists, at least in relation to human evolution). Modern evolutionary biologists concur that the emergence of complexity has been an important aspect of the evolutionary process. However, they generally object to the view that evolution can be characterized by any single, predominant trend. There have been many trends, and many reversals of different trends. For this reason, evolutionists generally prefer Darwin's definition of evolution as simply: "descent with modification." One can speak of different directional trends as different forms of "progresssive" evolution. There are also many examples of " regressive" evolution in these terms. But there is no such thing as "devolution". This is the terminology that I will employ here.

In any event, my theory is addressed to the phenomena that Spencer equated with evolution itself. It is about the evolution of organization in nature and human society. It is not another so-called "systems theory"; these "theories" have so far been largely descriptive and taxonomic. Rather, it is a theory **about** systems - a theory which seeks to explain **why** complex systems have evolved. What I will discuss here, however, is the aspect of my theory which is of most interest to anthropologists, economists, political scientists and sociologists: namely the emergence of complex socio-political organization in human societies. In particular, I will address the following "why" questions: Why, specifically, have socio-political processes arisen in the course of human evolution? Why have there been apparent direction trends over the past several thousand years? What are the necessary and sufficient conditions for such phenomena to exist? And what are the underlying sources of socio-political change - development, transformation, decay, and even collapse?

III. EVOLUTIONARY CAUSATION

The first of the three key elements of the theory involves, in effect, a reversal of our commonsense understanding of the nature of causation. In evolutionary processes, causation runs completely backwards from our intuitive understanding - and from conventional social science methodology. That is, effects are causes: It is the functional consequences, or **effects** of various phenomena that are the preeminent causes of systematic evolutionary changes (and continuities) over time - in conjunction with various stochastic and deterministic factors. This is not really a totally alien idea. It is also the basic causal relationship posited by behaviourist psychology (see Brown and Herrnstein, 1975), and it was given formal (verbal) expression by psychologist E. L. Thorndike back in 1911 and 1985 as the so-called Law of Effect (although Thorndike's law was anticipated by Darwin much earlier, see Candland **et. al.**: 1977, pp. 45-46).

In effect, then, natural selection is really a theory about functional effects as causes. Natural selection is not an active selecting agency or mechanism out there in the environment somewhere. It is a way of classifying those aspects of the dynamic, functional interactions within organisms, and between organisms and their environments, that are responsible in a given context for causing differential survival and reproduction among genetically different individuals. A nice illustration involves a study by Bryan Clarke and J. J. Murray, Jr. (Clarke, 1975) of micro-evolution in a population of the English land snail, **Cepaea nemoralis**, inhabiting a sand dune at Berrow England. When these snails were first studied by an earlier researcher in 1926, it was found that there were two different varieties with markedly different shell coloration. One variety was solid-colored (single-banded), while the other was striped (or double-banded). At the time, the double-banded variety was clearly predominant; it had greater reproductive fecundity. However, when the same population was re-surveyed in 1960 (12 snail generations later), it was found that there had been a radical change in the frequencies of the two varieties. Now the single-banded variety were predominant.

How come? What was the **cause** of this example of micro-evolution? In essence, it involved a complex set of changes in the total configuration of functional relationships. It happens that **C. nemoralis** are preyed upon by thrushes; the thrushes have developed the clever habit of breaking open their shells on rocks. In 1962, the amount of predation at Berrow was relatively low and the relative proportion of the two snail varieties was determined more by the superior fecundity of the double-banded variety. In the intervening years, however, the dune had been invaded by a shrub called the sea buckthorn, which provides shelter for the thrushes, which are themselves preyed upon by hawks. As a result of this change in the ecology, the population of thrushes increased, and so did the amount of predation on the snails. However, the birds preyed much more heavily upon the double-banded variety, because their shell markings were much more visible.

In other words, the "cause" of the change in the frequency of the two snail varieties was the net functional effects, or consequences of a whole chain of interacting changes in the relationships between the snails and their environment.

Now, this same kind of causal process also applies to socio-cultural evolution. In recent years, a number of social scientists have proposed that cultural evolution can be characterized by a form of selection (**e.g.**, Campbell, 1965; Corning, 1974; Durham, 1976; Boehm, 1978; Blute, 1979; Chase, 1980). I call it "teleonomic selection" because, in actuality, this form of selection is based on internal "purposes", or goals - i.e., needs and wants that express themselves in the form of specific motivational and response propensities. In a nutshell, I contend that it is the functional consequence (anticipated, previously experienced, or even observed) in relation to "revealed" needs or wants that is the underlying cause of systematic cultural shifts - in accordance with Thorndike's law.

A classic and well-documented example is the automobile. In the latter 19th Century, there were innumerable experiments with various road-vehicles. In fact, at the turn of the century there were a host of different types of vehicles - Stanley Steamers, three-wheeled electric cars that resemble somewhat today's powered wheel-chairs, vehicles that literally looked like horseless carriages (such as the Duryea), and so forth. Nevertheless, in 1900 there were still only about 10,000 powered vehicles in the United States, versus about 25 million horses.

Then along came a rapid increase in national wealth during World War I, plus the discovery of vast amounts of crude oil cracking processes which dropped the price dramatically, plus some significant improvements in the design of the automobile (such as the self-starter), and of course, Henry Ford's mass production techniques. The net result of this and many subsequent changes was that, by 1960 (or about 12 horse and automobile generations later), the population of horses had declined to less than 5 million, while the population of automobiles had surpassed 75 million.

Needless to say, I have glossed over many of the complexities associated with the operation of these two principles. Some of these complexities are discussed in Corning (1983). Suffice it to say here that the key to both natural selection and teleonomic selection as agencies of directional evolution is the production of new functional effects that canalize change by differentially rewarding the adoption, perpetuation or elimination of different organic or social processes and artifacts. And this brings us to the concept of functional synergism - to what I call the "synergism hypotheses".

IV. FUNCTIONAL SYNERGISM

The concept of synergism may already be familiar, in relation to such every-day phenomena as drug side-effects or corporate mergers. However, it

happens that synergism is far more fundamental and important than is
generally appreciated.

In brief, synergism means simply "to act together", or "co-operate". It
refers to combinatorial effects that cannot be produced by various
elements, or parts or components acting independently. It is what we mean
when we say that "the whole is greater than the sum of its parts" (see
Pantin, 1968). In the physical realm, synergism is manifest in every one
of the periodic table of elements, as well as in many chemical compounds.
For instance, the light metal sodium and the poisonous gas chlorine are
obnoxious to man, by themselves. But when the two elements are combined,
they produce a substance that is positively benificial to man (in moderate
amounts) - ordinary table salt. Likewise, when two organisms engage in
symbiotic cooperation, they may achieve adaptive advantages that are not
otherwise possible. A case in point is lichen - actually a partnership
between an algae and a fungus. Because the former is able to engage in
photo-synthesis while the latter has capabilities for water retention and
for gripping surfaces, the two together are able to inhabit many barren
environments in which they could not survive alone.

The same kind of synergistic, or "co-operative" effects apply to the
behavioral realm as well - from the many documented examples of symbiosis
between species to the co-operation between mating pairs in sexually-
reproducing species to the multi-faceted cooperative behaviors in **Apis
mellifera** (the true honey bee), involving such tasks as nest-building,
foraging, reproduction, defence, and environmental conditioning. (Many
specific examples are cited in Corning, 1983).

Does this same principle apply also to the evolution of human
societies? I discuss this issue at length in the monograph, but here I
will cite just one example - a famous one. In **The Wealth of Nations** (1976
(1776)), Adam Smith describes a pin factory that he personally visited in
which ten workers were together able to produce about 48,000 pins per day.
They were able to do so by dividing up the production process into a
number of component tasks, each of which lent itself to specialization.
However, if each laborer were to work alone and attempted to perform all
of the tasks associated with making pins, Smith doubted that on any given
day they would be able to produce even a single pin per man.

This is, of course, the textbook example of the "division of labor". However, the slogan "division of labor" obscures a deeper principle - that of functional synergism. Another way of looking at the same phenomenon is in terms of how various specialized skills and production operations were efficiently **combined** into an organized, goal-oriented "system". The system Adam Smith described included, not only the roles played by the workers, but also the appropriate machinery, energy to run the machinery, sources of raw materials, a supporting transportation system and "markets" - i.e., a way of selling what was produced to recover production costs and perhaps make a profit. (Smith was reputedly the first economist to recognize that economies of scale are limited by the extent of the market - by the number of customers who tacitly cooperate in sharing the production costs). Finally, the pin factory was dependent also upon the existence of a cybernetic (decision-making, communication and control) sub-system through which planning, the hiring and training of workers, production and work coordination, marketing, bookkeeping, **etc.**, were effectuated. In sum, the "economies", or positive synergism realized by Adam Smith's pin factory were the result of the total system - the total configuration of man-man, man-machine, man-society, and man-environment relationships. Remove any single component and the system would not work.

This example could be multiplied almost endlessly, for the principle is applicable also to hospitals, hotels, shopping centers, insurance companies, banks, buses, subways (when they run), airliners, universities (sometimes), telephone systems, postal systems, movies, television, newpapers, books, computers, space programs, labor unions, armies, even governments.

Indeed, some of our most commonplace institutions actually embody multiple forms of synergism. Consider the local restaurant. For starters it involves an example of the classic division of labor among the various specialized tasks (Maitred', chefs, bartenders, busboys, waitresses, dishwashers, cashiers, cleaners, bookkeepers, managers, **etc.**) (Whyte, 1949). In addition, there are the combinatorial effects associated with the total "dining experience" (the decor, the view, the seating arrangements, the table setting, the service, the food, the wine, the other diners and even the rest rooms). There is also synergism in the tacit cooperation through which the various patrons together share the total cost of supporting the restaurant and its operation. There is even nutritional synergism in the food combinations that, in a well-balanced meal, provide complementary sets of nutrients. Moreover, if the restaurant is located in a shopping center or is part of a well-know chain, it may benefit from being linked to various external organizational entities.

Now, the concept of functional synergism, whether in nature or in human societies, is not a new idea. But, to my knowledge, nobody has heretofore suggested that synergistic effects are also causal - and of fundamental importance. Let me state the hypothesis explicitly. To quote myself: "It is the selective advantages arising from various synergistic effects that constitute the underlying cause of the apparently orthogenetic (or directional) aspect of evolutionary history - **i.e.**, the progressive emergence of complex, hierarchically-organized systems."

I maintain that synergistic effects have been of central importance in the emergence of biological organization of all kinds. The principle is as applicable to the evolution of the eukaryotic cell and multicellular organisms as to colonial organisms, endosymbionts, exosymbionts, and animal and human societies - that is, to cellular, organismic and sociobiology, as well as to the social sciences. In short, the synergism hypothesis involves a unified theory. It is, if you will, a functional theory of biological and social structures.

I dare not dwell too long on this aspect. Let me just add the following points, without elaboration: First, the concept of functional synergism is neither tautologous nor vague. Quite the opposite. From a functional standpoint, synergistic effects are always **relational** - they are dependent upon the consequences in relation to the "state" of the system (specific goals, or needs in specific contexts). Thus, when I am hungry, an ice cream Sunday may be very nourishing. When I am full, it may give me acute indigestion. Indeed, there are also a great many examples of "dysergy", or negative synergism. Negative drug synergism is a case in point. And so are the Rube Goldberg inventions that gather dust in people's attics because they don't work, or don't do anything particularly useful. In fact, only about half of the patents granted in the U. S. are ever utilized, while less than 10 percent of all new products are marketed successfully and the vast majority of new businesses eventually fail (see Blute, 1979).

Furthermore, the functional consequences (and thus selective potential) of a particular synergistic combination very often depends upon the relative merits of other alternatives that may be available. There may be better or worse automobiles, restaurants, and pin factories (as we all know). In short, the synergism hypothesis conforms to the principles - and metrics - developed by the cousin science of ecology and economics. What is more, it **encompasses** these disciplines; it provides a unifying systems framework and a unifying functional principle. (Applicable formal models of mutualism, collective goods, and positive sum games are discussed in Corning, 1983).

Accordingly, synergism is an eminently measurable phenomenon. In fact, it is routinely measured by economists, ecologists, ecological anthropologists and others. It can be measured in terms of monetary costs (and benefits), or in terms of energy inputs and outputs, or time expended (or saved), or reproductive efficiency, or any other functionally significant quantity - the number of raspberry tarts produced per hour, or the number of customers that choose to walk into your restaurant, or the number of cars sold, or the number of riders that share that cost of supporting a transportation system, or the number of votes cast.

Among the many well-documented cases that are cited in the monograph, one of the most dramatic, surely, involves the "huddling" behaviour of emperor penguins. In order to survive the long and brutal antarctic winters - when temperatures can reach minus 60 F. and winds can attain 100 KPH - these remarkable animals congregate into tightly packed colonies numbering in the tens of thousands for several months at a time. In so doing, each animal shares precious body heat with others (and provides insulation), while receiving body heat and insulation in return. Individual energy expenditures for thermal regulation are thus reduced by up to 50 percent, according to estimates based on experimental data (Le Maho, 1977). Without the synergistic effects of such cooperative behaviour, Le Maho concluded, it is very doubtful that any of animals would survive the winter.

Another example, in the human societies, involves negative synergism. If a commuter train carrying 400 passengers to work is 15 minutes late due to a malfunction that would have required one man-hour of preventive maintenance to avoid, with the result that the passengers are all 15 minutes late getting to work, the combinatorial effect is a net loss of 99 man-hours of productivity.

A final point, before moving on, has to do with how competition and conflict fit into this hypothesis. Briefly, the synergism hypothesis is central to explaining a major paradox in social life. It is curious, when you think about it, that the most significant forms of competition and social conflict occur between organized, co-operating entities - football teams, tribal groupings, corporations, labor unions, political parties, revolutionary movements, armies, and even military alliances between nations. The paradox dissolves when the various organized forms of competition and conflict are reduced to examples of functional synergism in relation to specific social, economic or political goals. Simply put, cooperation for the purpose of engaging in competition or conflict is merely a sub-set of the total array of social phenomena in which synergistic effects are causally important. This is not to say that functional synergism provides a complete explanation for the causes of war or revolution, obviously. But it does explain why such conflictual processes are very often carried on via organized cooperation. Furthermore, synergistic effects may explain some of the causes, not to mention the outcomes. Indeed, the relative effectiveness in achieving the synergistic potentialities of an organization may have a decisive effect on outcomes. A good case in point in the commercial realm, I believe, is the recent success of the Japanese automobile manufacturers vis-a-vis the Americans (Vogel, 1980; Ouchi, 1981). Likewise, in Keith Otterbein's (1970) important study of the evolution of warfare, a strong correlation was found across a sample of 50 primitive societies between the level of "military sophistication" and "political centralization", on the hand, and both the frequency and degree of success in waging wars.

In other words cultural evolution - by extension political evolution as well - has really been a **dualistic** process, involving both mutualistic and competitive, or conflictual forms of cooperation. Yet, both facets of this process have shared two fundamental properties in common - one is functional synergism and the other is cybernetic communications and control processes. This brings me to the third element of the theory.

V. THE CYBERNETIC MODEL

I would hope that the cybernetic, or control-system model does not require
an elaborate introduction. I assume most social scientists have at least a
basic familiarity with cybernetic concepts from the work of Wiener (1948),
Ashby (1956), Deutsch (1963), Easton (1965a, 1965b), von Bertalanffy
(1973), Buckley (1967, 1968), Steinbrunner (1974), Coulam (1977), Miller
(1978), and others. I discuss the cybernetic model at some length in
Corning (1983, pp. 44-45, 192-215, 318-323). Here I will just draw
attention to certain key points: First, acceptance of the cybernetic model
has been retarded significantly by the fact that it is widely
misunderstood. The basis of the misunderstanding can be made clear by
comparing David Easton's version of the model (Figure 1) with that of
William T. Powers (1973a, 1973b; also 1978). Note that in both versions of
Easton's model the innards of the system are portrayed as being an empty
"black box". To Easton, politics is characterized as a conversion process
in which inputs (demands, supports, and feedback) are "converted" to
outputs. Easton's system is essentially teleological in character. Its
behaviour is slaved to exogenous goals. It is more like a thermostat than
a self-organizing biological system.

Now compare this with Powers' model (Figure 2). In this version of the
cybernetic model, a teleonomic, or internally goal-directed system is
posited. In this case, endogenous "goals", or "reference signals" are a
major determinant of system behavior. (Of course, these endogenous goals
may be in-built, or the result of developmental "programming", or
situationally-determined, or some combination of the three). However,
internal goals do not alone determine behaviour. Behaviour is actually **co-
determined** by a process of matching feedback from the system outputs to
its reference signals (or its internal "program"). Feedback is always a
function of the relationship between the system (including its "goals",
"needs", "wants") and its specific environment. (Feedback is, in fact, the
"mechanism" by which the Law of Effect operates, although that is not the
only function of feedback processes).

COMPLEX AND SIMPLIFIED VERSIONS OF EASTON'S MODEL

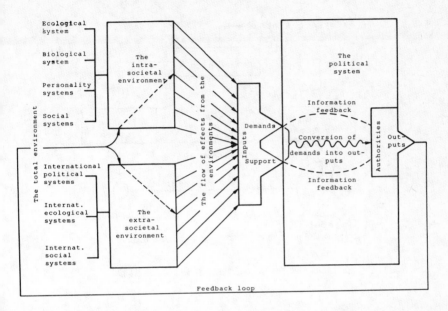

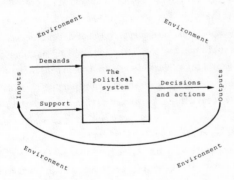

(From Easton 1965b)

FIG. 1

A CYBERNETIC CONTROL SYSTEM

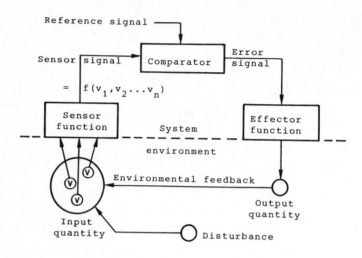

(After William T. Powers (1973))

Fig. 2

The secound point is this: Without going into detail here, I will simply assert my belief that the case for viewing the cybernetic model as a valid idealization (or simplified description) of real-world teleonomic systems - from eukaryotic cells to "individuals" to nation-states - has grown increasingly compelling. Every level in the organizational hierarchy of life has certain common, cybernetic properties: namely, goals, "decisions", goal-directed actions, communications and control processes. In Corning (1983), I attempt to sort out some of the subtle distinctions between the various so-called systems theorists, including those who accept most of the elements of the cybernetic model without calling it that, or those who do not posit goal-directedness explicitly yet develop models that require such an inference to be made (e.g., Miller's monumental, 1000-page disquisition, **Living Systems**, 1978, which is a kind of Gray's **Anatomy** of cybernetic systems).

Accordingly, the third point is that I maintain it is appropriate to define politics as the cybernetic aspect (or sub-system) of social systems, of all kinds. In my definition, families have political sub-systems (I differ with Aristoteles here), and so do football teams, symphony orchestras, and universities - and so do honey bees, wild dogs, chimpanzees, baboons, wolves and other social species. Furthermore, political systems may come into existence, develop, be transformed, fission, decay or even collapse. When the last bars of the symphony have died away and the orchestra goes home to bed, so does its political system. And the same is true of an intramural football team, and the Shah's government. Thus, the cybernetic model is **not** inherently either "radical" or "conservative". It is isomorphic with the dynamics of the real world.

A corollary of this point is that competition, conflict and the "struggle for power" are not the essence of politics but an aspect - one major class of goal-directed social behaviours manifested by individuals or organized political systems. There are, in principle, two other broad classes of political behaviours: (a) co-operative and (b) autonomous but "ecologically" interdependent (see below).

The final point, before turning to my theory of politics, is that if one accepts the cybernetic model of the "innards" of biological and social organization, it raises a major "why" issue - namely, why have cybernetic systems evolved in the course of evolution? Currently there is no "why" theory. Miller (1978), for instance, has generated many important cross-level hypotheses about the operation of living systems, but he does not offer an explanatory theory. He does not attempt specifically to account for the emergence and "progressive" evolution of living cybernetic systems. This is what my theory does attempt to do. And when the "why" question is focussed strictly on the social level of organization (as opposed to the cellular or organismic), then the issue becomes identical with the problem of explaining the emergence of politics (as here defined). So, let me turn to that issue.

VI. A GENERAL THEORY OF POLITICS

To repeat, my theory of politics is an integral part of the theory of "progressive" evolution defined above. The basic postulate is that functional synergism has played the key role in the directional evolution of man and society toward more complex forms of social organization. In accordance with the model of causation that I described earlier, selection based on synergistic effects of various kinds has been the underlying cause.

However, at every step this progressive trend has been **co-determined** by the development of more elaborate forms of cybernetic regulation and control. Beginning with the most elemental of life processes, in fact, cybernetic control processes are what have made goal-oriented biological (or biosocial) processes possible. And, as each new level of biological organization has evolved, there has been a concomitant "progressive" differentiation - via natural selection and/or teleonomic selection - of specialized mechanisms of decision-making, communications and control.

Though I cannot elaborate in detail here, let me very briefly enumerate some of the propositions and predictions that flow from this basic vision. (There are 12 included in Corning (1983), and more that could be adduced). The first two propositions involve what I call the "Iron Laws of Organization"

1. **Who says a division of labor, or a combining of functions (cooperation), at any biological level, says organization.**

2. **Who says organization, says cybernetic control processes (i.e., political processes, or their functional equivalent).**

It is not coincidental that these laws are a paraphrase of Roberto Michels' "Iron Law of Oligarchy" (1949 (1911)), for there are both similarities and differences between his formulation and my own. First the similarities. Contrary to the interpretations of various critics, Michels' law was, like my own, based on an essentially functional argument. A brief (two-page) chapter in his classic work **Political Parties** was devoted to the "technical" and "administrative" factors that necessitate organization. "Be the claims economic or be they political, organization appears the only means for the creation of a collective will. Organization, based as it is upon the principle of least effort, that is to say, upon the greatest possible economy of energy, is the weapon of the weak in their struggle with the strong" (1949 (1911), p. 25).

Up to this point in his argument, I am basically in agreement (though there is much more to organizational synergism than energetic efficiency), and our laws might very well lead to similar predictions regarding the emergence of political systems. Indeed, it is often overlooked that Michels correctly predicted before the fact the political - i.e., organizational - course of events that would follow various 20th Century socialist revolutions. "Socialism is also an administrative problem", he pointed out (1949 (1911), p. 403).

I go beyond Michels, though, in viewing the functional significance of organization much more broadly and in focussing explicitly on cybernetic processes. My theory is concerned with cybernetic functions in relation to the production of all manner of synergistic effects. Furthermore, it adheres strictly to a functional perspective and takes no **a priori** position on whose goals may be embodied in the system. These laws could apply equally well to "rule over" as to "rule on behalf of" the parts, or members. They apply to slave systems and to utopian communes alike.

In contrast, Michels assumed the worst and held that there is a built-in tendency toward oligarchical and exploitative rulership. This, he argued, is due to various inherent inequalities (inherited, experiential and/or cultural), coupled with "psychological factors" (especially nepotism). Michels was not breaking any new ground in this respect, of course. Similar arguments can be found in Plato's **Republic**, Aristotle's **The Politics**, Machiavelli's **The Prince** and **The Federalist**, to name a few. However, in my theory I leave it an open question whether or not a given system is effectively realizing its intended (explicit or implicit) goals, whatever they may be. A system may be more or less successful, or efficient, in attaining its goals ("eufunctional" or dysfunctional", in functionalist jargon). Moreover, the precise "mix" in a given system between (1) the personal values, objectives and skills of various political actors, (2) the constraining or facilitating influence of the "political culture", and (3) various structural ("engineering") characteristics of the system (inclusive of reciprocal "feedback" controls, in cybernetics terms) can only be determined by analysing the specific case in point.

In any event, I believe that these Iron Laws of Organization are eminently falsifiable in principle, though I hold that it cannot be done. In fact, these laws cannot be falsified in any other social species, or at any other level of biological organization, so far as I know. (This critical issue is discussed further in Corning, 1983).

Assuming, then, that a deterministic linkage does in fact exist between politics **qua** social cybernetics and social organization, this leads to a pair of subsidiary propositions, one of which involves a falsifiable prediction - namely:

3. Political (cybernetic) processes are a requisite for the realization of functional synergism in social organization

Therefore,

4. Given sufficient "capabilities", any failure to achieve designed or "intended" goals will be due to specific failures of the relevant cybernetic communications and control processes.

I contend that the historical record (and the literature in political science, sociology, economics, and management science) is littered with supporting evidence - case-studies of political failures, from lost elections to lost revolutions to lost wars, and from defeated football teams to bankrupt organizations. Indeed, this proposition implies that many of the internal problems of organizations that are conventionally treated under separate categories - **e.g.**, leadership, morale, management practices, communications, training, **etc.** - can be reduced to facets of the problem of cybernetic (political) control. The critical test, then, is the following: The proposition can be falsified if any instance can be found where, despite sufficient capabilities, nevertheless there was a failure minimally to attain organizational goals for reasons **other** than political.

It may already have occurred to the reader that there are **many** apparently falsifying exceptions to this proposition. "Accidents" do happen, after all. However, I contend that these are, on second thought, exceptions that prove the rule. Admittedly, in many cases organizational failures are the result, not of a lack of adequate resources or poor organization in a structural sense, but of unexpected and perhaps unpredictable (stochastic) factors - the wind changes that led to the defeat of the Spanish Armada; the unusually early and severe winter that stalled, fatally, the advance of Hitler's armies into Russia; the iceberg that sank the Titanic; the fire that consumed the Hindenburg; the Iranian-hostage rescue fiasco; the new technologies that suddenly make an old technology obsolete.

Nevertheless, such manifest failures of prediction can also be reduced to failures in cybernetic terms and can be incorporated into the scope of the above proposition. Cybernetic control is always a function, not of some ideal of perfect knowledge, but of whatever information happens to be available to the system (built-in information, information inputs and feedback). By definition, cybernetic processes entail dynamic control of the relationship between a system and its environment. This is why insufficient information, uncertainty and risk-taking are endemic features of real-world decision-making and organizational management (Simon, 1957). Accordingly, an organizational failure due to unpredictable events constitutes one type, among others, of control failures on the part of the system. It can properly be said that a system is poorly designed, or functionally "maladaptive", if it (or its operators) is unable to predict and compensate for perturbations that result in failures to achieve the operative goals. In other words, insufficient information is quintessentially a cybernetic problem. Indeed, the very definitions of what constitute "information" and a "sufficiency" of information depend on the nature of the goals and the specific context; that is, there is no (biological) information that does not actually or potentially relate to goals. The rest is only **potential** information, or noise.

Let me provide one concrete example. It involves two hospitals that were recently studied in depth by a management consulting firm. These two hospitals - let us call them hospital A and hospital B - are so strikingly comparable to one another in most respects that they come about as close to being a controlled social experiment as one is likely to find, ready made.

Both hospitals are privately owned, non-profit institutions operated by religious orders. Both are located in the same state (and are therefore subject to the same regulatory environment). Both are located in urban, residential, low-income minority areas. Both are of comparable size - about 180 and 230 beds, respectively. Both have about the same mix of patients, in terms of reimbursement for services.

Nevertheless, hospital A produces a consistent 6-8% net return as a percentage of gross revenues, while hospital B has averaged less than 1% for the past five years and in 1979 ran at a loss. The explanation, as far as can be determined, does not lie in any of the "externals". In fact, if anything hospital A is at a significant disadvantage. It has an older physical plant (10 years versus one years); it has a slightly higher load of charity and low-reimbursement Medicaid patients; it has a high volume of maternity patients (a traditional "loss-leader" in the hospital business), whereas hospital B has none at all; it also has 50 fewer beds. Even more significant, hospital A has double-occupancy rooms, whereas hospital B has only single-occupancy - which is far easier (at least in theory) to keep full. Hospital A has the problem of matching different sexes, matching smokers and non-smokers, and so forth.

The differences in performance outcomes between the two hospitals, in sum, can, be traced to their political (cybernetic) systems. Hospital A has a management system that is a model of how to run an institution, while hospital B provides a model of how **not** to run an institution. In hospital A, the director is open, supportive, sensitive to his staff and to the needs of the institution. He is also decisive, well-organized and committed to the development of the hospital. Hospital A also has a formal planning and control system that runs throughout the organization. Planning is thus an integrated, system-wide process. Each unit participates directly in developing the overall hospital goals and is "brought aboard", so that everyone is working toward the same goals. Management also helps promote the feeling of a collective effort. Furthermore, short-term goals and activities are developed in an annual M.B.O. (Management by Objectives) plan. These plans are in turn related to longer-range goals. Thus, the hospital staff does not sabotage objectives but lines up behind them.

In addition, the **external** relations of the hospital are open and participatory. The board of trustees is not dominated by the religious order but includes members of the medical staff, the hospital staff and the community. Members of the community also serve on every one of the hospital's boards and committees.

The result: Hospital A runs very lean. It is a highly efficient operation, with tight cost-controls. Patient volume is also an incredibly

high 95% of the maximum occupancy rate for its medical and surgical
services, despite the double occupancy. One can even "feel" the difference
in the hospital environment. The place is spotless. Staff morale is high.
The staff is friendly and cooperative but business-like. People are busy.
Nobody stands around killing time.

Contrast this with hospital B, where the director is young,
authoritarian and closed. He hands down orders, rather than consulting. He
is also ambitious to move up in the corporate hierarchy of the religious
order, and so calculates his actions accordingly. In addition, there is no
formal planning process at all in this hospital. The board of trustees is
drawn exclusively from the religious order, and all are from out of town.
Planning, when it is done, is completely **ad hoc** and often done at cross-
purposes. As an example, one committee got the hospital to commit itself
to renovating some medical office space. When the job was half done,
another committee persuaded the hospital to shut it down altogether.

The result: This hospital runs a very lax operation. Productivity is
low; there is a "chit-chat culture" and many operating inefficiencies.
Cost-controls are minimal. Patient volume, moreover, is only 75 - 80 % of
maximum occupancy, **despite** the hospital's single-occupancy room
arrangements. Hospital B's expenses per patient day run $ 483, versus $
459 for hospital A, which amounts to $ 1.57 million per year more in
operating expenses.

The conclusion seems well founded: Given similar resources and
capabilities, the differences in performance between these two hospitals
are the result of factors in their **political systems** (in the sense in
which I use the term here and in the monograph).

A second pair of subsidiary propostions are related directly to the
issue of political evolution. The first propostion states:

5. **The greater the degree of organization complexity, the greater
the need for commensurate political qua cybernetic processes.**

This proposition in turn leads to a prediction:

6. **While there may be some short-term "lead" or "lag", over the
longer-term there will always be a high degree of concordance
between the level of complexity attained by any social system as
a whole and the complexity of this political (cybernetic) sub-
system.**

In Corning (1983), I discuss the existing theories that relate to these propositions and review the evidence in some detail. It spans the literature in organization theory, small group theory, communications theory, cultural anthropology and political science. Here, let me simply illustrate with some cross-cultural data. First, there is Carneiro's (1967) comparison of 46 single-community societies, to determine the relationship between size (as measured by population) and the number of organizational traits. As Figure 3 indicates, Carneiro was able to plot a regression line such that the number of organizational traits approximated the square-root of the population (or $N = .6P^{.594}$, to be precise). Even more compelling were Carneiro's (1970) correlations between seven categories of cultural traits (including political and legal) based on the ethnographic literature for a sample of 100 societies. These correlations, which are highly significant, are shown in Table I.

Of course, these data relate to the structural (and functional) **differentiation** of a society. But there are also data relating to structural **integration.** Using Murdock and Provost's (1973) set of 10 societal development scales for a sample of 186 societies, I was able to obtain Spearman rank-order correlations between their scale for political integration and each of the other nine scales (with the lowest correlation being $r_s = .346$ for the scale relating to modes of transportation). All the correlations were significant at the p .001 level.

Rolf Wirsing's (1973) study of "political power" (**i.e.**, control over decisions in five public domains) across a sample of 25 pre-industrial societies was also highly supportive. Wirsing developed a set of scales for decisions regarding: (1) the initiation or prevention of warfare, (2) adjudication of disputes, (3) appointments to lower offices, (4) allocations of wealth, and (5) assignments to involuntary labor. He found that the degree of political power was functionally related to the number of "organizational levels" on which politically-relevant information was stored and retrieved. Equally important, he found that power distributions were much less closely related to population size per se. In effect, Wirsing's results supported the hyothesis that political power is strongly associated with the ability to control access to, and distribution of, information, and that hierarchical organization fosters power concentration.

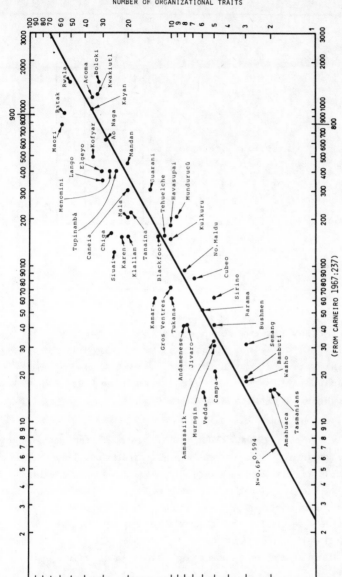

NUMBER OF ORGANIZATIONAL TRAITS

POPULATION SIZE AND SOCIAL COMPLEXITY (N=46)

POPULATION

(FROM CARNEIRO 1967:237)

FIG. 3

not used

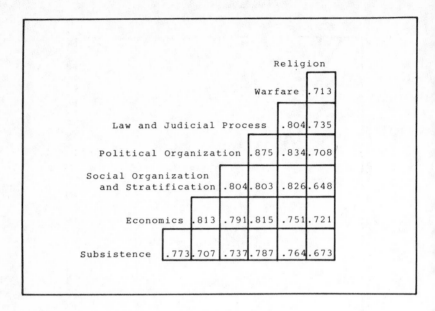

                                          Religion

                              Warfare   .713

      Law and Judicial Process   .804 .735

        Political Organization .875 .834 .708

    Social Organization
       and Stratification .804 .803 .826 .648

              Economics .813 .791 .815 .751 .721

    Subsistence .773 .707 .737 .787 .764 .673

TABLE I

Correlations between seven categories of culture (n=100).
(From R. L. Carneiro, "Scale Analysis, Evolutionary Sequences
and the Rating of Cultures", in R. Naroll and R. Cohen, eds.,
'A Handbook of Method in Cultural Anthropology' (Garden City,
N.Y.: Natural History Press, 1970).)

Similarly, McNett (1970a, 1970b; 1973), in a factor analysis of 89
selected cultures, found extremely high correlations between his rank-
order scale of settlement pattern "types" and the characteristics of a
society's economic, political and religious institutions. Equally
important, various kinship, household and other "social" traits were **not**
significantly correlated. McNett concluded that his data strongly support
the functionalist view that there is a close relationship between the
areal distribution of human populations and basic, pan-societal
institutions.

Perhaps the single most compelling cross-cultural study, however, is
the highly innovative multi-factor analysis by Alan Lomax (with Norman
Berkowitz) (1972), which involved a pair of new methodologies called

cantometrics and choreometrics - the quantitative analysis of song and dance styles and performance patterns. Because these expressive and communicative aspects of cultural evolution have been shown to correlate highly with evolutionary changes in economic and political systems, and because song and dance patterns are functionally related to the reinforcement of societal cohesion and organization, Lomax has found that cantometric and choreometric "profiles" can serve as sensitive indicators of cultural patterns in general and communications patterns in particular. Using a factor analysis for 71 variables (including many conventional economic and political measures) across 148 of the cultures listed in the **Ethnographic Atlas,** Lomax was able to "reduce" the central characteristics of cultural evolution to a single global attribute - increasing organizational capabilities for controlling and manipulating the relationship to the environment. This "grand scheme" in turn was found to involve a complex interplay between two "deep attributes", "differentiation" and "integration".

There are also a number of studies showing parallel trends within major macro-level political "sub-systems" - legal, administrative, military, etc. For instance, Carneiro's findings (cited above) regarding the correlation between legal complexity and various other culural categories was buttressed by the studies undertaken by Schwartz and Miller (1970) and Wimberly (1973) in which legal complexity scales were developed in relation to the progressive evolution of such specialized legal roles as mediators, courts, police and "counselors" (or advocates). Similarly, as noted above, Otterbein's (1970) study showed that there is a close correlation between the level of overall political centralization in a society and the degree of specialization and technological sophistication in the military sphere.

In a major review of some 150 cross-cultural studies, Raoul Naroll (1970) lists as "historically valid" in "broad outline" the following findings regarding cultural evolution (among others): (1) There has been a clear-cut, allometric trend toward greater occupational specialization (excepting for a major "spurt" since the industrial revolution); (2) a corollary has been a trend toward ever-greater accumulation of information and informational technology; (3) a similar trend can be observed in the evolution of more diverse and complex task-oriented organizational types,

or "team" types in Naroll's term; (4) several factor analyses have been
consistent in showing a pronounced historical trend in political system
complexity and authority patterns - from egalitarian, informal, non-
coercive and of limited scope (functionally and structurally) to
hierarchical, formalized, coercive (via **both** rewards and sanctions) and
multi-functional in scope. The "progressive" evolution of political
systems, in sum, has been an integral part of the larger, systemic process
by which culture has evolved.

My fourth pair of propositions relate to the causal role of politics in
socio-cultural evolution. I posit that functional synergism and cybernetic
processes are **co-determinants** of cultural evolution - the **stable**
"progressive" political developments will always be accompanied by
positive synergism, and vice versa. (Recall the example of Adam Smith's
pin factory). Furthermore, the initiative for such progressive changes can
come from political ("governmental") sub-system, independently of
functional changes in the "economic" system (broadly defined), though
often **both** kinds of changes occur, together.

Here I definitely part company with the Marxists. First, I dispute
Marx's view that politics is epiphenomenal and lies outside of the mode of
production. Rather, politics **qua** social cybernetics is an integral part of
it. Second, I attempt to show that politics is neither exclusively a
dependent variable nor an independent variable. It can be both. In the
full monograph, I cite some specific examples from the literature on
cultural evolution - namely the rise of the Zulu nation and the adoption
of wheat cultivation by the Pima Indians (Corning, 1983, pp. 365-367). I
also relate this view of cultural evolution to the debate concerning the
origin of the state (Corning, 1983, pp. 368-375) which I view as a
dualistic process that has involved cybernetic concomitants of both
"economic" and military forms of co-operation. Finally, I suggest that
there may be considerable "play" in the relationship between the economic
aspect (broadly defined) and the political aspect of social life. This is
illustrated in Figure 4, which combines a modified version of Gerhard
Lenski's (1970) economically-oriented taxonomy of societal "types" (after
Goldschmidt (1959) with Kent Flannery's (1972) politically-oriented
taxonomy after Service (1962).

THE "PLAY" IN POLITICAL EVOLUTION

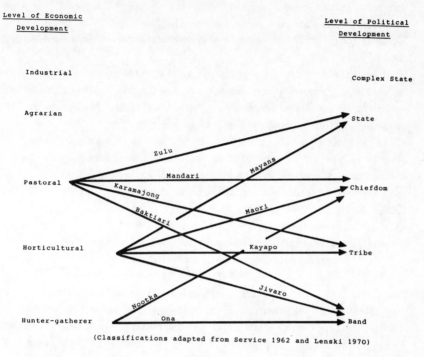

(Classifications adapted from Service 1962 and Lenski 1970)

FIG. 4

Several observations are in order with regard to this illustration. First, such taxonomies of economic and political "types" tend to emphasize (sometimes unduly) a particular aspect of a systemic process. They also may obscure many subtle similarities across categories and many differences within categories. (In fact, Lenski argues for differentiating between simple and advanced forms of horticultural and pastoral societies on just such grounds). For instance, the Maya can be classed as horticultural in terms of their basic food production technology but they were also highly developed in terms of various crafts industries and were distinctive among horticulturalists in evolving a "state" form of political organization. Furthermore, many societies are mixed types; still others may not remain fixed in a single category but may evolve over time or even manifest cyclical patterns. The bison-hunting Indians of the North American plains, for example, underwent seasonal aggregations from composite bands into the equivalent of chiefdomships in response to the seasonal patterns of the animals they exploited for their food (Oliver, 1974).

Nevertheless, there are significant limits to the amount of "play" that is possible. Complex states do not arise on a horticultural base, and agricultural societies are never organized as simple bands. Within the range of possible economic and political types, moreover, there does tend to be a clustering. In the 1966 edition of the **Etnographic Atlas,** 90 percent of the 147 hunter-gatherer societies listed were nomadic bands, whereas only 4 percent of 377 horticulturalists were nomadic. Most were quasi-sedentary tribes and chiefdomships. Furthermore, Lenski (1970, p. 130), in an admittedly preliminary analysis, determined that the median size of hunter-gatherer groups is about 40 persons (n=62), whereas the median size of horticulturalist groups is about 95 persons (n=45).

I will mention only one more proposition here. It has to do with political "regression" - simplification, fissioning, decay or collapse (Corning, 1983, pp. 375-382). The Marxian dialectic emphasizes progress via conflict. In my theory, •by contrast, societal change is viewed as having involved **both** progressive and regressive evolution. Political systems can and do disaggregate, either by mutual consent (when the symphony orchestra goes home or the political campaign is terminated), or by coercion. Come to think of it, the Marxian dialectic cannot even conceive of a civil war that leads to disaggregation - such as the fissioning of the Roman Empire or the dual division of the Indian subcontinent, first into India and Pakistan and later into Pakistan and Bangladesh. Neither can it handle the documented trend in socialist and capitalist countries alike toward **more** not less politics. (The rise of a **de facto** opposition labor party in Poland is, of course, a mockery of Marxist theory, as was the Iranian revolution). In other words, what the Marxists must do hand-stands to rationalize, the cybernetic theory explicitly predicts. The proposition states:

> **9. Political "regression" - the simplification, dismemberment or collapse of cybernetic social processes - is always associated with a loss of functional synergism.**

This is really the obverse of Propositions 7 and 8, alluded to above. It should be noted at the outset, however, that the loss of functional synergism may not be a "bad" thing. As noted, political systems are often deliberately, willingly disbanded. This occurs when the goals of the participants (or at least the controllers) have been achieved - when the hunt has been completed and the bison have been consumed, or when the buzzer, bell, horn or whistle announces the end of the game, or when the last bars of the symphony (or rock concert) have died away, or when the last hurrahs have been sounded in a political campaign and the candidates retire to await the returns, or when Johnny comes marching home again. If there are "costs" associated with sustaining any goal-oriented system, it makes no economic sense to maintain it when there are no longer offsetting benefits.

Of course, we do not always dismember an organization whose task has been completed. Consider, for instance, those obscure government commissions that continue to live on as middle-class make-work, or perhaps to satisfy some "latent function" (after Robert Merton) - say as a lightning-rod in case of trouble. Or consider The March of Dimes   - the classic case of a volunteer organization whose original mission (fighting polio) was spectacularly accomplished. The March of Dimes has found other battles to fight and has become The March of Megabucks.

However, the cases in which we are most interested, and the ones that are the most challenging theoretically, are the systems whose goals are ongoing, or still to be attained in the future - the systems in which regressive changes are **imposed** upon the system, so to speak. When a family undergoes a divorce, when a company goes bankrupt (at least in the old days, before the new bankruptcy laws), or when an empire declines and falls, its political system also disaggregates. Sometimes such changes involve analogues of what biologists call "adaptive simplification" - that is, changes that are actually beneficial for the goals of the system (or its "parts"). When a system is re-structured so as to eliminate deadwood, remove operating inefficiencies or otherwise improve its capacity to perform its mission, structural simplification may have positive functional concequences. This is not inconsistent with the proposition above. When more inclusive systems no longer produce positive synergism (when the costs outweigh the benefits, or there is "dysergy"), a regression may enable the system, or some of its parts, to continue operating "profitably". The Chrysler Corporation provides an obvious current example of adaptive simplification. At this writing, the outcome is still in doubt. But there is certainly no question about the intention behind Chrysler's corporate weight-reduction program.

By the same token, adaptive simplification can also involve a fissioning process among incompatible "parts" whose combinatorial interactions are unprofitable, on balance. An example close to home is the Stanford Research Institute, which was founded and grew up inside of Stanford University. In time the different missions of the university and what is now known as SRI International became increasingly inharmonious, and the institute was spun off as an entirely separate entity.

There are also many examples of macro-level political regression, needless to say. The premier modern cases are various 19th Century European empires, but there are also many more remote cases and countless more or less well-documented cases in the archeological literature (see Eisenstadt (1963), Wesson (1967), and Taagepera (1968, 1978a,1978b, 1979, 1981)). In fact, so far as I know, there is not a single extant case of an autonomous state-level political system with an unbroken history going back more than a few hundred years, a very short time in evolutionary terms. Both historical and prehistoric states have frequently been subject to disaggregation or, sometimes, absorption into more inclusive entities. The reasons, ultimately, have to do with changes or breakdowns in the synergistic combinations of factors that make such growth processes possible in the first-place (inclusive of political factors, needless to say).

By far the best documented and perhaps most illuminating historical example is, of course, the Roman Empire. The Roman case illustrates well the proposition above, for recent scholarship has shed new light on the relationship between the Roman economy and its political fortunes. What has recently become more apparent is the fact that, throughout the entire cycle of Roman history, there was always an intimate relationship between functional synergism (both economic and military) and the fortunes of the Roman polity, both when it was in the ascendancy and when it was in decline. Indeed, the lessons seem directly pertinent to our present historical circumstances. (A brief overview is provided in Corning (1983, pp. 377-380).

VII. SOME THEORETICAL IMPLICATIONS

Finally, let me summarize very briefly how, in my view, this theory relates to two of the major theoretical domains of political science - political development and international relations. (In the full monograph, I also discuss revolution and war). With regard to "development", first, it might be said that political development is to political evolution as ontogeny is to phylogeny. Though developmental processes may recapitulate past evolution, either through borrowing or through independant invention, they are not the same thing. Political development involves the establishment of a cybernetic infrastructure for any goal-oriented social system; it occurs every time a viable organization comes into being. Political evolution, by contrast, occurs only when a **novel** function is adopted and when appropriate new political structures are created, or when a novel organizational entity is created to perform an existing function. Likewiese, "progressive" evolution occurs only when such novel functions entail the elaboration of a more complex political system, or when an existing function is subdivided and handled through more complex structures. (Political "regression" involves exactly the converse situation, of course). Thus, all viable polities develop; some also evolve. But in all cases, the same principles and propositions that were articulated above will apply.

As I see it, developmental theorists have often conflated these two different things, and unless I am much mistaken, the distinction I have drawn cuts through one of the most pervasive sources of confusion in political development theory. Political development involves the organizational "capabilities" for achieving social goals, whatever they may be. It does not necessarily imply greater complexity, modernization, democratic institutions, or what have you. Many functionally-oriented writers on political development during the 1960s and 1970s, taking their cue from Gabriel Almond's seminal formulation (Almond and Coleman, 1960), have also recognized the broader process of political evolution. As Almond has observed: "The generic themes of cultural secularization, structural differentiation and specialization, and increasing capacity or performance ran through all of these writings " (1973, p. 5). One might even be able to make the argument that, all other things being equal, "modern" political institution and democratic practices are on the whole more effective means for achieving certain political goals and functions. However, this conclusion does not follow as a logical necessity, nor is it true in all cases (or for all political goals). The resort to "emergency government" in wartime is a case in point. The so-called "Lincoln dictatorship" at the outset of the American Civil War, for example, was a rational and necessary response to this situation at hand, given Lincoln's objective of saving the Union. (see Rossiter, 1948)

Accordingly, while it is not possible to say that a given polity is **under-evolved**, it is possible to say that it is **functionally under-developed**, in the sense of **not having the requisite institutional structures and functional capabilities for the achievement of its stated, or tacit goals.** Many precariously integrated Developing Countries today would fall into this category. But consider also the case of the United States at the turn of the century. Though more highly evolved than some European states of that time in terms of citizen participation, competitive parties and other "feedback" processes, it was comparatively under-developed with respect to various regulatory, military and social welfare functions (and attendant organizational capabilities).

Conversely, it may be said that some polities are over-developed: Their (cybernetic) superstructures are more elaborate, and/or more controlling than is warranted by the needs and objectives at hand, with possibly detrimental results. Such a condition could readily be ascribed to certain "command" economies, though it could also be said that these countries are over-developed in some respects and under-developed in others. Opinions might be more divided concerning the issue of over-development in some of the so-called "post-industrial" welfare-states, such as the United States. (I will sidestep that currently lively debate). On the other hand, it is permissible to say that an evolutionary change might be desirable, beneficial, or even a requisite for the achievement of a particular goal. The prospective international ocean-bed regime is a perfect example, but so was the U. S. space program.

The converse of political development, then, is a decline (or collapse) in terms of functional efficacy. To the extent that Samuel P. Huntington (1965, 1968) employs his well-known concept of political "decay" as an efficiency criterion, this is in accord with my definition. However, political decay is not (functionally) equivalent to regressive evolution (or to "devolution", in some anthropological formulations). Nor is it equivalent to the abandonment of democratic political practices, however much we may deplore such an eventuality on moral grounds - or on more far-seeing functional grounds, for that matter. Huntington (and others) use the term decay in all of these functionally inconsistent ways.

It should also be emphasized that the distinction I make between political development and political evolution does not mean that two processes are necessarily mutually exclusive. Far from it. The cloning of state governments during the westward expansion of the United States and the establishment of various colonial governments by 19th Century imperialist powers constituted both political development (at one level) and political evolution (at a higher level). Both processes occurred simultaneously. Furthermore, evolutionary innovations occurred at both levels over time.

By the same token, one can observe both developmental and evolutionary processes at work in the international arena today, with the Developing Countries striving to create viable national institutions that are more or less closely modelled after existing patterns, even as historically novel processes of national **and** transnational political evolution are occurring. This is not to say that the Third World polities are perfect replicas of industrialized polities. To the contrary, they are invariably composites of unique cultural (and historical) circumstances, elements borrowed from other cultures and political innovations of their own making - just like the Developed Countries, when one thinks about it.

So why do states develop, and evolve? In short, because of historically determined, situation-specific **opportunities** for achieving functional synergism in relation to specific "internal" and/or "external" social goals. As I argue at some length in the monograph, these goals are generally (though not always) derived from "needs" that are related more or less directly to the survival and continuity of a polity and its members. (See also the earlier discussions of this point in Corning (1971, 1975) and the similar argument by Waltz (1979 pp. 91 - 92)). However, a linkage must also be made between opportunities, goals, and actions.

The opportunities (capabilities plus rewards) for development/evolution arise out of the specific context. The Roman Empire was built with a

combination of natural resources, a strong agricultural base, a
technically superior crafts industry, opportunities for economical
shipborne commerce around the rim of the Mediterranean, seige-engines and
military phalanxes armed with iron swords, shields and body armor, horse-
drawn chariots and wagons, good roads, aquaducts, writing and record
keeping, legal and administrative mechanisms and a host of other factors.
Roman hegemony was not the result of a . preordained plan but of a
piecemeal, "organic" growth process - a favorable tide that spanned
several centuries. Likewise, the nation-states that emerged in Europe from
the 16th Century forward were favoured by complex combinations of
ecological, technological, economic, socio-political and military factors.
Take away any major element - say artillery - and Europe might have
remained a fragmented patchwork of warring Feudal baronies that were
protected, ultimately, by impregnable fortresses (Andreski, 1968 (1954)).
Such evolutionary changes in turn create new needs and new opportunities.
It is an endless process.

From Shakespeare to Arthur Schlesinger Sr. (1939), to Karl Deutsch
(1979), historically-minded observers have often used the metaphor of the
ocean tides to express this idea; political evolution occurs when
historical opportunities are "taken at the flood". The precise
configuration of circumstances is always unique, to be sure, but the
synergism principle and political processes are the underlying constants.

The second domain that I will briefly discuss - international
relations - manifests rather less coherence in terms of perceptions among
various theorists about the basic "problem", and unit of analysis. In a
major review of the field in 1975, Kenneth Waltz characterized the state
of I. R. theory as "weak and confused". Research on the subject has been
of low fertility, nothing seems to have accumulated and there have been
few major gains, Waltz maintained. More serious, Waltz perceived deep
division of opinion over what constitutes the basic nature of the subject
matter - how it is to be apprehended, what phenomena are to be included,
who the principle actors are and how they relate to one another. There is
also disagreement, Waltz noted, about what it is that a theory of
international relations should explain. What is the dependent variable?

Then there is the school of thought which holds that international
relations is not amenable to the "scientific method" at all - that it is

intractably historical and particularistic. Dougherty and Pfaltzgraff, while trying to be upbeat in their comprehensive survey of I. R., nevertheless concede that, on the "crucial question" of wheter or not a scientific theory is possible: "The authors cannot answer that question with any certitude" (1971, p. 38) (cf. Zinnes, 1975).

Thus, the field seems at present to be fractionated among a variety of narrow-gauge general "paradigms", on the one hand, and various constructs that are oriented to middle-range theoretical problems. Following (roughly) the taxonomy developed by Dougherty and Pfaltzgraff, one might tentatively identify the major "schools" of international relations as (1) Realists (including "balance of power" theorists), (2) Systems Theorists, (3) Interdependence and Integration Theorists, (4) Conflict Theorists (inclusive of war, revolution, imperialism and nuclear deterrence), (5) Decision-Making Theorists, and (6) Quantitative Theorists.

These schools cannot be discussed here in more detail. Suffice it to say here that the various "approaches" are not necessarily antagonistic to each other and may more easily be reconciled than one might suppose. For instance, Waltz makes a persuasive argument for the view that imperialism, while a very real phenomenon, is not peculiar to capitalist societies (**viz.**, the invasion of Afghanistan). In fact it has been ubiquitous among tribal societies, chiefdomships and pristine early states (Corning 1983). Imperialism (or better said, political expansionism) has been an endemic aspect of what Waltz aptly characterizes as a "self-help" system. It is likely to occur whenever there are (1) overwhelming military and/or economic power imbalances between independent "actors" (inclusive of differences based on size) **and** (2) sufficient positive incentives (**net** benefits). Such political acquisitiveness has, in fact, played a major role in socio-political evolution generally.

In effect, my own approach to a theory of international politics takes off from Waltz's reconstruction effort (1979). Waltz argues that, if **Realpolitik** and balance of power theories capture the essentially "anarchic" and competitive nature of international politics (he defines the word anarchy carefully), these theories "fare badly" in predicting the course of events because they are **too** reductionalist, for the most part. That is, they do not take sufficient account of the "system" in which each

of the actors is embedded. By "system", Waltz does not mean that
international politics consists of a single unified entity; it is not
analogous to an organism (**i.e.**, a cybernetic system) that is endowed with
an overarching goal (or goals) and a functional division of labour, but
rather to a "market". As Waltz points out, a market is viewed by economic
theorists as a structure, or framework of forces with emergent properties
(synergistic properties in my term) that arise from the relationships and
interactions among the actors. Let me use Waltz's words here:

> The market arises out of the activities of separate units - persons
> and firms - whose aims and efforts are directed not toward creating
> an order but rather toward fulfilling their own internally defined
> interests by whatever means they can muster. The individual unit
> acts for itself. From the coaction of like units emerges a
> structure that affects and constrains all of them. Once formed, a
> market becomes a force in itself, and a force that the constitutive
> units acting singly of in small numbers cannot control. Instead, in
> lesser or greater degree as market conditions vary, the creators
> become the creatures of the market that their activity gave rise to
> ... (Thus), the market is a cause interposed between the economic
> actors and the results they produce. It conditions their
> calculations, their behaviors, and their interactions. It is not an
> agent in the sense of A being the agent that produces outcome X.
> Rather it is a structural cause. A market constrains the units that
> comprise it from taking certain actions and disposes them toward
> taking others.

Up to this point, I second Waltz's line of reasoning. The chief difficulty
with various so-called systems theories of international politics, such as
Morton Kaplan's (1957), is that they fail to make a clear distinction
between the two fundamentally different kinds of systems in international
politics - which interact with each other. Waltz calls them "firms" and
"markets", but I would call them "cybernetic systems" and "ecosystems". I
differ from Waltz in that I advocate the explicit use of the biological
cum ecological model, in **lieu** of the economic model. The difference is, I
believe, significant.

Though the market model is certainly more familiar to social
scientists, in point of fact the organism-ecosystem - or ecological
community - model provides a far richer and more precise analogy with the
actual behavior of both economic systems and international politics.
First, the market model defines too narrowly the goals of the actors.
Neither in international politics nor in economic life do the actors
single-mindedly pursue only "power", or "profits". At the minimum, they

pursue their survival   - which, on reflection, is not a simple goal. There is no such thing as "mere" survival. Survival is always a continuing and multi-faceted challenge (Corning 1971, 1975, 1981). However, the actors may also pursue a wide variety of goals that are instrumentally related to such things as the growth, power and prestige of the polity, or organization. Even the profit motive turns out to be complex, when one begins to probe the motivations behind it (see Marshall, 1890).

Second, neither in international politics nor in economic life do the actors pursue their goals only by "struggle", or "competition". They do so by a variety of means and with a variety of relationships to other actors. Besides competition, there are such ecological analogues as competitive exclusion, predation, parasitism, reciprocity, mutualistic symbiosis, niche partitioning, ecological specialization and multifarious forms of inadvertent ecological interdependency (see Odum and Odum, 1971 (1953), Krebs and Davies, 1978, and Pianka, 1978). There is even occasional altruism.

Third, and perhaps most important, political and economic activities are, in the aggregate, related to each other and to the biological problem of survival and reproduction for human populations. Indeed, our political and economic systems also interact with natural ecosystems and affect them to such a degree that "ecological problems" are now a major agenda item in international politics (see Pirages, 1978, **inter alia**).

A classic rendering of the ecological paradigm is Darwin's description of the English countryside in the concluding passages of **The Origin of Species**. He wrote (in part): "It is interesting to contemplate an entangled bank, clothed with many plants of many kinds, with birds singing on the bushes, with various insects flitting about, and with worms crawling through the damp earth, and to reflect that these elaborately constructed forms, so different from each other, and dependent on each other in so complex a manner, have all been produced by laws acting around us. These laws, taken in the largest sense, being Growth with Reproduction; Inheritance....; Variability....; a Struggle for Life and.... Natural Selection" (1968 (1859), p. 459).

Accordingly, international relations (or the interactions among primitive societies and various historical states, for that matter), can best be studied as "political ecology" (see Russett, 1967, Sprout and

Sprout, 1965, 1968). Political ecology involves the interaction among
various goal-oriented, cybernetic systems in specific political, economic
and biophysical enviroments. These "super-organism" (in Herbert Spencer's
term) have various goals, various capabilities, and operate under various
kinds of constraints. They include nation-states, military alliances,
ethnic groups, multi-national corporations, common markets, terrorist
groups, international agencies, and all the rest. The relationships among
these actors are shaped to a significant degree by the political goals and
military and economic capabilities of the major nation-states, but the
political ecosystem includes many other actors in a variety of
relationships. It includes economic and political cooperation **and** conflict
of various kinds. (I develop the term "egoistic cooperation" at some
length in Corning, 1983, pp. 103 - 104, 255 - 258, as a way of
differentiating between those forms of cooperation that are self-serving
and those that entail altruism; see also Axelrod, 1981; Axelrod and
Hamilton, 1981; Hirshleifer, 1978; Ghiselin, 1974; 1978; Wilson, 1975,
where the relations between cooperation and concurrence are discussed in
more detail).

In short, the overarching paradigm of a community of goal-oriented
cybernetic systems embedded in a political ecosystem (or systems) can
encompass the full range of motives, actors and relationships that exist
in international politics. More important, political ecology provides a
paradigm within which functional synergism can be observed to play a
central part. The synergism hypothesis can account **in part** for the various
goals, strategies and actions that are found in the international arena -
from the nation-states themselves to multinational corporations to U. N.
agencies. It can, for example, help us to understand the sources and
changes in power "capabilities", as well as some of the breakdowns that
have occurred in the balance of power. The patterns of war, conquest and
hegemony among organized polities are also illuminated, and so too are the
patterns of coalition and military alliance. (These subjects are discussed
in more detail in the monograph, with particular reference to recent
quantitative modelling and political economy approaches).

Equally important, the synergism hypothesis help to explain the myriad
forms of international economic and social cooperation - sometimes in ways
that transcend military alignment patterns (such as Soviet purchases of U.

S. wheat, or U. S.-Soviet scientific collaboration). It also helps us to understand the **dualistic** historical trend toward greater war-making potential coupled with greater economic interdependence and political integration - from primitive bands to nation-states to various transnational entities.

In addition, the synergism hypothesis can enable us to make more sense of many of the seemingly irrational, contradictory and even bizarre patterns of political aggregation and disaggregation; it makes them consistent and, to a limited degree, even predictable. Thus, it is not at all sursprising that the United States might enter the Vietnam war in part to check what was perceived to be the "domino" effect of Chinese expansionism and emerge from the war moving toward a tacit political alliance with the very same nation that had recently been calling the United States a "paper tiger", that was abhorrent for its atheistic communism, and that had actually fought against the U. S. in Korea. Indeed, the Vietnam war itself, perverse as it may seem, may have contributed to this meeting of minds. In the wake of the Vietnam disaster, our relative losses, Russia's relative gains and China's more clearly evident weakness combined to make such an alliance seem far more advantageous to the U. S.. If the two nations "act together" they may be able to check Soviet expansionism, so the argument runs.

This is, of course, only one of the more dramatic and immediate examples among many that could be cited. The proposition also enables us to understand why the U. S. and Russia became allies in World War Two, despite the earlier U. S. intervention in their civil war and our implacable antagonism to communism, as well as why the two countries might become arch-enemies after the war, when their common enemies had been annihilated. One can speculate that, had that war ended in an early armistice rather than in "unconditional surrender", we might still be allied with the Russians against the "axis" powers. Similarly, it explains why the British and French could fight one another for more than 1000 years and then suddenly do a turnabout and join forces to fight the nascent German state in two of the greatest human slaughters of all times.

These examples could be multiplied almost endlessly. In all such cases, the "higher" morality of human needs (as well as more crass forms of self-interest, to be sure) fairly consistently override political ideology and other value-orientations. When there are significant opportunities for achieving functional synergism ("economic" or "military") through political integration (cybernation), these may override less important sources of political cleavage. Though we may (or may not) decry this state of affairs, we can predict that such supposed political contradictions will be commonplace. (For a "functionalist" approach that might be viewed as a prologomenon to my theory, at least insofar as it relates to international politics, see Haas (1964). The recent volume by Keohane and Nye (1977), is also compatible. Other relevant works are Sewell (1966); Haas (1970); and Mitrany (1975)).

But can we go any further? Surveying the current scene, can we venture some "hard" predictions as well? What will be the future course of political evolution? Will there be a nuclear shoot-out and a return to a stone-age culture? Or will "organic" processes of sociocultural and economic evolution lead to greater political integration - to regional polities or a world-order.

Like Darwin's theory (properly understood), this theory makes only one unequivocal prediction - that the future cannot be predicted with certainty, because it is not a linear, deterministic process but a highly contingent concatenation of "chance, necessity and teleonomy" (to embellish the slogan of the late Jacques Monod). The future depends upon changes in our natural environment that cannot, at least now, be foretold with any real confidence. It also depends upon constraints we may not as yet even comprehend, plus ideas, invented and other developments (new forms of functional synergism) that are inherently beyond our ability to predict. Most importantly, the future will also depend upon potentially decisive political actions. (Recall again that functional synergism is always relational and situation-specific; it is dependent upon the "fit" between all of the different elements and processes in the system, and between the system and its environment).

Some contemporary cultural evolutionists are more sanguine about predicting the future. Naroll (1967), for instance, analyses the historical trend in Eurasian empires toward ever-larger aggregates and concludes with a range of probabilities for achieving a unified world state that ranges from the year 2125 (with a 40 % probability) to 2750

(with a 95 % probability) to 3750 with a probalility of .99935. In the
meantime, Naroll believes wars between currently independent political
units will continue to occur without significant abatement. Carneiro
(1978) generally concurs, though he employs a different methodology for
making his own prognostications. He projects forward the historical trend
toward a decreasing number of politically autonomous units. Provided that
nuclear annihilation is avoided, he asserts, "a world state cannot be far
off" (1978, p. 219). It is a matter of centuries or even decades, and not
millennia, he claims.

Thus, we have come full circle back to Spencer's Universal Law of
Evolution. Though the social sciences owe more to Spencer than is
generally acknowledged, in this fundamental respect we would do better to
cast our lot with Darwin; I believe we have just as much to learn about
the future from the extinction of the dinosaurs as we do from studying our
own past history as a species. It is presumptuous of us to think that,
even now, we have that much knowledge of, or teleonomic control over the
evolutionary process - that we have transcended "chance and necessity".

But if we cannot make deterministic predictions, we can make many
conditional, "if-then" predictions. For example, in due course it is quite
likely that there will be dramatic changes in the structure of the nation-
state. The nation-state is not vouchsafed by any law of nature, after all;
it is an historical artifact associated with a particular sociocultural
context. It could be that the nation-state will eventually be replaced by
a more inclusive political entity (or entities), or that they will regress
into less inclusive entities. Or, paradoxical as it may seem, it could
also be that the modern state is evolving in both directions at once.

Some more inclusive integrative functions might in time be
"surrendered" upward - either through an armed conflict the scenario and
outcome of which boggle the imagination - or through a piecemeal extension
of the global social contract beyond postal unions and such like to, say,
the international monetary system, transnational labour conditions, the
oceans, aspects of environmental regulation, and so forth. Certainly, if
contemporary economic trends continue, the augury is for more, rather than
less, political integration in due course. In whatever manner such a
centripetal movement might occur, it can be predicted that it will only
endure as long as it remains, on balance, functionally advantageous.

Conversely, if there is a significant diminution in some of the forces that were responsible in the first place for integrating the sometimes strange bedfellows that comprise modern states, it can be predicted that there will also be **centrifugal** tendencies. Both progressive and regressive changes might occur simultaneously - depending, as always, upon their functional consequences. Only the scenario of apolitical socialism is ruled out.

By the same token, failure to solve what John Platt some years ago called our "crisis of crises" - the nexus of problems involving population, food, water, energy, resources and the environment - will in the long run have catastrophic (or at least degenerative) consequences. For better or worse, we are dependent upon enormously complex systems of co-determining "variables". And if the Club of Rome's famous "limits to growth" projections grossly oversimplified the situation, nevertheless they contained an element of truth. A point that is emphasized in the monograph is that synergistic combinations also create dependencies. If the requisites to sustain a system are for any reason undependable, the seeds of ultimate collapse may be sown at the very outset. However, foresight has not been the rule, either in nature or in human societies. We need only look at the record.

It should be stressed once again, though, that functional consequences are not wholly determining of evolutionary change. Men do not always seize the potential opportunities that are available to them, nor do they always act expeditiously to break up political systems that are dysfunctional. The patterns conform to the "quasi-rational choice" model described in Corning (1983, pp. 244-245). Opportunities for functional synergism must first be discerned, discovered or "invented" - and communicated. There must also be a cognitive alignment of perceptions regarding functional needs and the functional consequences of various actions, an alignment that is often impeded by perceptual filters of various kinds. Furthermore, these perceptions must lead to decisions and goal-directed actions.

The metaphor that comes to mind traces to the historical roots of cybernetics, in World War II. The dynamics of political evolution, both progressive and regressive, resembles the classic shipboard fire-control problem of tracking a rapidly moving target (an attacking enemy aircraft) with an anti-aircraft gun which also happens to be on a moving platform. As long as the trajectories of the ship and the target remain the same, the gun will eventually lock-in on the target and anticipate its position. But if either the ship or the target change course and speed, which both may be doing rather frantically in a battle situation, the fire-control problem becomes immensely more complicated; there must be constant, necessarily "lagged" responses to feedback and to new trajectory information. The "target" for human populations is, of course, "survival" in a complex world of cooperation and competition, but the same basic, cybernetic model applies. I believe that we will find not better way of modelling and understanding political processes than in such terms.

VIII.   CONCLUSION

What I have proposed is a nested set of theories about evolution as a unified historical process. Human evolution, and the changes we have wrought among ourselves (and in our environment), is, to be sure, **sui generis.** But this is a truism that, by itself, restricts our ability truly to understand man and society. The fact is that every morphological and behavioural invention and every "progressive" development throughout the past 3.5 billion years or so has been **sui generis.** Moreover, these inventions have been cumulative and are linked to one another through an organizational hierarchy of awesome architectural and functional complexity. At each new level of biological organization, new properties have emerged. Every new level of "wholeness" has not only transcended but, more important, ultimately **organized** and integrated lower-level parts.

Yet, at the same time, the parts always constrain and impose imperatives on the whole. Furthermore, at every emergent new level of biological organization, including human polities, this multi-layered process has manifested certain common properties. Progressive evolution of the kind that has concerned us here, whether at the microscopic level of one-celled creatures or at the most inclusive macroscopic level of transnational political organizations, embodies cybernetic processes coupled with functional synergism. Human political systems are not, therefore, a thing apart but a variation (and elaboration) upon one of the most primordial of evolutionary themes. Not only do the purposes served by

our political systems trace by indirection to our basic needs as
biological creatures, but the patterns of organization that we have
evolved are successful only insofar as they embody the same basic
principles that govern lower levels of biological organization.
Furthermore, just as natural selection, through trial and success, has
oriented the evolutionary process toward more "progressive" forms of
biological organization, so teleonomic selection has oriented
sociocultural evolution toward more "progressive", i.e., complex, forms of
political organization.

Like Darwin's theory, this theory does not attempt to account in detail
for every specific historical event. Again, the causes always involve
unique sets of historical factors. This theory suggests, rather, what the
general form of the explanation will be for certain "progressive" (and
"regressive") aspects of the evolutionary process - what general
principles will be operative and where to look for both past and **future**
"progressive" developments (if they occur). In other words, this theory
does not imply an end to social sciencies but rather a new approach to
organizing and integrating our data - an approach which crosses a number
of disciplinary lines. Furthermore, this theory makes no promises about
the future, though it does define the nature of the problem and the
general nature of the solution. Indeed, it views the future as one of
unending challenge - of great uncertainties but also of great
opportunities. It also views politics - at all levels of society - as the
very heart and soul of the process of human adaptation, both as
individuals and as collectivities with shared, "public" purposes. Politics
is ultimately about our survival, and the survival of our species.

I believe this theory can more than satisfy the criteria I set forth at
the beginning for what I conceive to be a general (and scientifically
defensible) theory of politics. I fully expect sharp criticism of any
weaknesses. I also anticipate the necessity for making refinements, in
accordance with the original meaning of a dialectical process as a way of
gaining understanding. However, I also believe that this theory is
fundamentally sound, and that it will withstand the test of time.

BERNHARD GIESEN

MEDIA AND MARKETS

I. INTRODUCTION

The revival of sociological evolutionism during the sixties has largely ignored the rise of neo-Darwinian theories of evolution in biology and related fields of research; instead of utilizing the conceptual refinements and theoretical achievements of neo-Darwinian theorizing it fell back upon the classical paradigm of growth and unilinear development. This evolutionist paradigm depicts social change along the line of increasing differentiation, rationalization and adaptability (the functionalist version) or according to the irresistable logic of stages of moral and cognitive development (the Habermasian version). The evolution of modern society is viewed as closely tied to the institutionalization of specific **media of interaction.**

Starting with some critical comments on the methodology of classical evolutionsism an alternative theoretical model based on the concept of 'selection' will be presented and applied to the analysis of 'media of interaction'. Thereafter the focus will be on the specific situation conditions and categorial presuppositions which enforce and enhance the application of 'media codes' whereas other situational conditions (also present in modern societies) inhibit and counteract the operating mode of the media codes. Instead of presenting modernity as an irresistable and global development of macro structures the evolution of modern codes of interaction will be described as depending on contingent and varying situational structures.

M. Schmid and F. M. Wuketits (eds.), Evolutionary Theory in Social Science, 171–194.
© 1987 by D. Reidel Publishing Company.

172   B. GIESEN

II. THE SELECTIONIST PROGRAM

The idea of "growth" implied in most functionalistic conceptions of social
change is closely connected with a "categorial-analytic" strategy of
theorizing; this strategy aims at aprioristic models reconstructing the
"rationality" of institutions, world-views, historical developments and
patterns of action. Its philosophical core consists of the Kantian theory
of knowledge and the Hegelian conception of history (Münch, 1981; Münch,
1982, p. 17). From this point of view history as well as empirical reality
are considered as fields of analytical reconstruction, instead of an
independent basis on which critical arguments against the theory itself
can be grounded. This "categorial-analytical" model of theorizing can be
contrasted with a critical alternative presenting a different conception
of the relationship between theory and experience, history and evolution,
situation and rationality. When explaining social actions this alternative
does not focus on the rational constructions and plans of the actor, but
on the empirical condition and situational presuppositions of the action;
when reconstructing historical change it does not stress the growth of
complexitiy and rationality but the contingencies affecting this growth,
for example scarce resources and stable or turbulent environments; when
analyzing world views or systems of norms and rules it does not insist on
ordering cognitive structures according to the logic of evolutionary
stages but centres on the conditions of institutionalization and social
conflict; when accounting for the relationship between theory and
empirical evidence it does not conceive this relationship to be a matter
of analytical reconstruction but as the basis of mutual criticism.

I will call this alternative strategy of theorizing the "selectionist"
programme (Schmid, 1982c, pp. 176-209; Schmid, this volume; Giesen, 1983,
pp. 230-254; Giesen, forthcoming). Critical arguments against a
categorial-analytical theory frequently rely on some kind of selectionist
reasoning - even if this criticism is put forward by advocates of a
different categorial-analytical theory who point out the complexity of
social facts and historical processes (Habermas, 1981, vol. 2. pp. 384-
419). With respect to the theory of evolution the contrast between these
theoretical programmes has been referred to by Sahlins as the difference
between theories of general evolution and theories of specific evolution

(Sahlins and Service, 1962, pp. 13, 69; for a criticism of theories of general evolution Giesen and Lau, 1981, pp. 229-256).

Selectionist arguments can be used as an occasional criticism against the lofty ambitions of categorial theorizing but it can also be enlarged and systematically incorporated into a general evolution-theoretical model that thereby escapes the pitfalls of naive progressism and is able as well to account for the requirement of a theory-based and non-arbitrary relationship between history and evolution (Nisbet, 1969; Granovetter, 1970; Smith, 1973; Schmid, 1982c; Popper, vol. 1, 1962).

While theorists in biology as well as in psychology, economics and philosophy of science are familiar with the application of these neo-Darwinian models of evolution, sociologists have paid little attention to the potential of neo-Darwinian reasoning for theories of social interaction or macrosocial change.[1]

An appropriate starting point for a Darwinian analysis of social interaction is provided by differentiating between the processes of **construction, selection and reproduction.** (Giesen and Lau, 1981, pp. 232-234; Boulding, 1978, p. 104). **Construction** refers to all those social processes by which individual or "collective" actors produce actions on the basis of rules, knowledge and patterns of typification. These rules may claim categorial or hypothetical validity, may be formally institutionalized or may be part of everyday knowledge: They represent the "symbolic structure" of action and serve as "blueprints" for the construction of action. This "constructive" function of rules and cognitive structures represents the main focus of the "categorial-analytical" strategy of theorizing.

Selection constitutes the indispensable counterpart of 'construction' but is largely ignored by analytical theories. Processes of selection are generated by problems of scarcity: the demands and goals of empirical systems mostly exceed the available resources. This holds for material as well as for social resources. The claims on validity which are associated with cognitive and normative propositions, and their rules and patterns of interpretations, exceed the actual scope of their empirical validity; they are subject to selection not only on the part of the actor, who reinforces the claim of a particular rule while neglecting or ignoring others, but also on the part of the empirical social situation. A claim on validity raised by an actor can be accepted and consented to by other actors or it can be rejected by them. Rules can establish a meaningful connection between different and subsequent actions or they can fail to do so. A normative orientation can successfully provide a basis of integration or it can generate conflict and dissent. In principle every process of action and interaction is based on a selection between the competing cognitive and normative orientations claiming validity. What determines the selective success of a particular rule or cognitive element with respect to its competitors are those **situational** conditions which are - at least partly - ignored and neglected by the actors. The analysis of these neglected but effective conditions of action and interaction is at the core of the heuristics involved in the selectionist programme.

Most sociologists oriented to empirical research would probably agree to distinguish between the claim on validity on the one hand and the factual or 'situational' realization of this claim on the other. It is however doubtful whether the evolutionary success of cognitive or normative elements can be estimated on the basis of a general standard and without any reference to the characteristics of a given situation. The Holy Grail of categorial-analytic evolutionism - growth of complexity and differentiation - fails to provide a workable criterion for factual selective success (Smith, 1973, pp. 60-95): While rare exceptions in the world of **symbolic** systems, de-differentiation and decrease of complexitiy are common phenomena on the level of **institutionalized** social rules; the mere knowledge of once obligatory rules (**e.g.** the complex rules of 'courtoisie' at the French royal court) does not imply continuing normative validity and social acceptance of these rules. This hints at the theoretical risk associated with confining the theory of social evolution to the analysis of normative and cognitive structures: the lack of a theory of institutionalization becomes preeminent.

Processes of institutionalization can be conceptualized as selective **reproductions** of normative and cognitive elements with respect to varying empirical situations. In contrast to the biological mode, the social mode of reproduction consists of particular procedures of **interaction** that transmit a normative or cognitive element from one actor to another: exchange, authoritative command and discourse are the different and alternative procedures of social reproduction by which actors attempt to move others to accept and adopt rules and cognitive patterns. I will refer to these procedures of transmission and reproduction as "codes of reproduction". These codes of reproduction are part of the pool of cognitive and normative elements representing the "genetic" or "constructive" basis of interaction.

In the same way as the selective success of rules and cognitive elements will depend on the given situation to which they are applied, the operative success of these codes of reproduction cannot be assessed without reference to the particular situation framing and conditioning the reproduction of normative and cognitive elements via social interaction. The characteristics of the situation can foster and inhibit, stimulate and restrict the application of codes of reproduction; not only the elements which are reproduced but also the codes by which they are reproduced are subject to the selective pressures of the situation.

One of the variables determining the "success" of a code of reproduction is the number of actors who are to be addressed by an "interaction offer". If there are only a small number of actors and time is abundant, even a very "slow" code of reproduction - **e.g.** discourse aiming at convincing individuals - is sufficient to address all individuals available for interaction. In this case codes which operate more rapidly would not provide any selective advantages. If however the number of possible 'Alters' is very large or even unlimited, a change to an accelerated code of reproduction will result in significant advantages: discursive modes of reproduction for example are too long-drawn-out, unstable and time-consuming to move a society to accept or reject a proposed normative innovation. These situations require 'rapid' codes like formal elections or authoritative commandments. On the other hand these 'rapid' codes may prove to be inadequate to situations including only a few persons: if somebody attempts to rely on legal claims when interacting with close friends he will generate bewilderment and indignation.

The following chapters focus on a special and advanced code of reproduction: social media of interaction. Before the situational preconditions for the successful operation of media code are analyzed the claim to validity implied by media of interaction has to be clarified: selectionist theories obviously presuppose the analysis of the elements which are subject to the selection processes.

III. MONEY AND LANGUAGE: TWO MODELS FOR GENERAL MEDIA OF INTERACTION

The notion of 'media of interaction' can look back at an impressive
conceptual career during the past decade. Originally used by Parsons only
for the social media: money, power, influence, and value-commitment, the
concept was extended later on to the subsystems of action (definition of
the situation, affect, personal capacity to act, intelligence) and finally
transmitted to the subsystems of the human condition (transcendental
ordering, meaning, health, empirical ordering) (Parsons, 1969, pp. 405-
438, 353-404, 439-472, Parsons, 1977b, pp. 204-228; Parsons, 1978, pp.
352-433; Parsons and Smelser, 1956).

When it comes to interpreting and conceptually clarifying the notion of
media with respect to sociological theory two different positions can be
assumed. The first position contends that media of exchange have to be
considered as **empirical institutions** which operate analogously to money.
Debates based on this position focus on the question whether media of
"interchange" can cope with various and diverse structures of interaction
or whether certain domains of interaction hinder any attempt to control
them by media because of their particular structure. Viewed from this
perspective media are particular codes for interconnecting actions.

In contrast to this the second position conceives of any social
specification of the meaning of interactions as media. Here the concept of
media takes on a very broad and unhistorical scope. In principle every
code specifying meaning can be conceptualized in terms of media: art,
love, friendship, law, solidarity, prestige, kinship **etc.** (Luhmann, 1975,
pp. 170-192). This conception models the functioning of media not on
economic exchanges controlled by money but on the structure of language
and the process of communication; it does not aim at describing an
empirical institution but at analysing any nexus of actions generated by
the specification of meaning. Whereas the 'money model' is based on
quantification the 'language model' does not confine media to this very
particular way of specifying meaning. Here media presuppose merely some
functional point of reference reducing the complexity and contingency of
meaning.

A "middle" position between the 'money model' and the 'language model'
is represented by the Parsonian conception wherein media serve as a kind

of "map" depicting possible ways of modernization. Modern societies have
institutionalized media only in some of their subsystems but future
thrusts toward modernization will make good this lag.

Any conception of media - whether it be the model of money or the model
of a functionally specified language - is based on the idea of a **symbolic-
symbiotical** code reducing and abridging processes of verbal communication
by reference to a specified function. Media are languages which are
confined to a single dimension of meaning but allow for a precise
representation of differences with respect to this dimension. In the same
way a linguistic statement represents a propositional content, a media
"statement" stands for social claims to opportunities of action: Money
stands for opportunities to exchange goods and services, power for control
over binding decisions, prestige for opportunities to be addressed by
gestures of esteem and admiration. Media codes imply a self-evident
structure of preferences; the desirability of money, power or prestige do
not need any particular justification. Therefore Ego can control Alter's
follow-up actions by 'staking' an appropriate amount of media. Problems
will not arise until the actors realize that the symbolic media are no
longer covered by the "assets" represented by them. A feeling of mismatch
between media and the assets or resources they represent will result in an
accelerated rise in the "price" to be "paid" for a particular asset or
resource. This inflationary dynamics of media - well known in the realm of
money - can be prevented by material or social guarantees to cover the
claims endorsed by media.

Such talk of the 'inflationary dynamics' of media puts us in mind of
the 'money model' (Baum, 1976, pp. 579-608; Gould, 1976, pp. 493-506;
Münch, 1976, pp. 135-149). Indeed, this money   model and related
statements of economics seem to provide a more detailed account of the
process of media controlled interaction than the model of 'language';
therefore the following analysis refers mainly to the conception of 'media
of exchange' or in the Parsonian phrase 'media of interchange'.

IV. THE INSTITUTIONALIZATION OF THE MEDIA CODES: STRUCTURAL REQUIREMENTS

The institutional core of media can be characterized by three
requirements:

(1) An institutionalized **standard** (ideally metric) for measuring social claims must be available. If this standard is lacking, the comparison of offers of interaction with respect to specific functions becomes extraordinarily difficult.

(2) The claims symbolized by the media have to allow for being **circulated** and **deposited**. Otherwise ascribing a claim to individual actors will be difficult.

(3) Finally the relationship between symbolic media and claims on actions or goods must be subject to generally accepted **control**. Without such an impersonal **guarantee of coverage** the application of any such code is limited to relationships of trust between personal acquaintances. Therefore, love is certainly a code of interaction but could hardly be described as a medium.

As indicated earlier it may be considered doubtful whether the institutional structure of media can be transfered to non-economic domains.[2]

If the media codes are not universally applicable to any kind of interaction, the **conditions** and **presuppositions** for their application and operation have to be established. Habermas has attempted to characterize these conditions. He names functional specificity, instrumental orientation and the possibility of binary decisions for those addressed by the media codes (Habermas, 1981, vol. 2, p. 395).

In making this qualification Habermas focuses on the **symbolic structure** of the actor's orientation. However the question may not be limited only to the symbolic or categorial structures presupposed by an action; it can also be extended to the **empirical** situation which frames and conditions the interactions controlled by the media codes. With regard to this empirical situation three theses will be proposed:

(1) The successful application of media to the control of interaction processes presupposes a "**market-type**" situation.

(2) This market-based context of institutionalization has to be distinguished from the **problems of interaction** the application of media is intended to solve. Therefore media can also be employed **a priori** to tackle non-economical problems.

(3) However the structure of the context of interaction on the one hand and the problems of interaction on the other **empirically** limit the control of interaction by media.

As indicated above the theoretical model framing and patterning the theses cited here assumes that the connection of actions to ordered interactions is based on symbolic codes the successful application of which depends on the empirical structure of a given situation. Of course this situational structure is itself generated by social processes. However with respect to the interaction under consideration it has to be regarded as a non-arbitrary, empirical frame which may not be questioned nor redefined by individual actors in the situation. The empirical structure of a situation is determined by fundamental differentiations of a social system, by the history of a relationship between the actors, by the continuity, stability and affectivity of this relationship and by the number of actors who may be addressed by an interaction offer.

This application of codes is basically restricted by two **limiting** structures: the categorial or **symbolic structures** patterning the orientation of the actors towards the situation on the one hand and the **empirical** structure of the situation conditioning the relationship between the actors on the other. This leads to a reformulation of the problem of institutionalizing. Neither the empirical structure of the situation nor the symbolic structure of the actors' orientations alone will provide a satisfying account of the process of institutionalization[3], but the relationship of matching and coping between these structures will.

V. COMMUNITIES, HIERARCHIES AND MARKETS

Three different types of symbolic and empirical structures need to be distinguished with respect to the successful application of codes of interaction: communities, hierarchies and markets. Each of these structural types corresponds to a particular code of interaction.

(a) **Communities** presuppose and emphasize the cognitive differentiation between inside and outside, between strangers and members of the societal communities. Their binary code is based on nominal concepts and aims at demarcating exclusive membership. The enclosed structure of preferences is obvious: the decision goes against the stranger. The more extensive and comprehensive the obligations of solidarity within the community, the more marked and insurmountable are the boundaries facing outsiders. By delimiting the realm of solidarity and increasing its obligation the number of possible holders of claims on solidarity are systematically restricted. At the same time personal ties of affection are generated between the members of a community; relations of solidarity between these members are symmetrical, incompatible with hierarchies and override possible differences in power. In several respects, relations of solidarity differ from the Habermasian normal situation. Ego cannot arbitrarily control his membership in communities and his corresponding obligations of solidarity. Interactions are not founded upon instrumental orientations of the actors. Ties of solidarity cannot easily be circulated or deposited if community membership is based on personal attributes which are not alienable or negotiable. The more the context of interaction approaches and resembles the structure of solidarity (demarcation between members and strangers, affectively grounded ties of solidarity between the members) the more the establishment of media will produce critical problems of interaction. This holds true even and especially with respect to functionally differentiated relationships of solidarity **e.g.** friendship and love in modern societies. Also the differentiation and graduation of solidarity tends to encourage crisisprone conditions with regard to ties of solidarity; it gives way to an unlimited enlargement of the community whose far-reaching borderlines are no longer demarcated by affective-cathectic differences, but by cognitive criteria. Without additional support by other institutions the code of community **inclusion** fails to provide an adequate structuring of the situation. For controlling relationships between spatially and socially distant actors the code of community inclusion has to be supported by a particular symbolism of membership.

(b) The symbolic structures presupposed by **hierarchies** retain the distinction between inside and outside but add to this horizontal difference the vertical differentiation between social ranks, between 'upside' and 'downside'. The actors' orientations towards each other differ with respect to their affective elements: feelings of solidarity between equals, hostility or respect toward superiors, contempt or indifference towards inferiors. The asymmetrical structure of hierarchies sets limits to the flexibility of relationships between individuals: persons of inferior rank usually cannot arbitrarily choose the ones whose commands they wish to obey and persons of superior rank only expect to

have particular commands obeyed. Functionally differentiated hierarchies
in modern organizations exemplify these restrictions. Here a claim on
dominance is referred to by a functionally specific code guiding the
decision of the lower ranks: legitimacy. In principle Ego can certainly
control Alter's decision to obey by increasing his threats; but this path
is limited by legal norms. Social hierarchies cannot be conceived of
without referring to these **normative** restrictions for the application of
power and force. They set limits to the degrees of freedom for choosing an
Alter for interaction as well as for choosing a way of controlling his
actions. In this regard the opportunities provided by hierarchies
certainly surpass the ones available in situations of the "solidarity"
type but they are confined to meeting legal claims which cannot be
arbitrarily augmented by Ego. Changing these normative controls over
actions usually requires cumbersome and time-consuming procedures as far
as large hierarchical organizations are concerned. This inertia enhances
the stability of organizations and strengthens their basis of solidarity.
Although hierarchically structured interaction certainly does not focus on
solidarity, solidarity is never completely eliminated and excluded as a
basis of interaction which may provide a foundation for legitimizing
claims and norms. As soon as the question of legitimacy is posed, the
verticality of the orientation is removed and the situational structure
changes from hierarchy to a communal or discursive pattern.

Stability of hierarchies is increased and enhanced if an additional
cognitive differentiation emerges: the difference between social validity
and individual opinion. As soon as the pressure to conform to social rules
does not rule out individuals having deviating opinions, and conformity
does not automatically involve individual adherence, the hierarchical
control of individual actions is more readily tolerated. However this
differentiation also gives way to instrumental attitudes towards
conformity and to the development of utilitarian foundations of morality
and social integration.

(c) Situations of the 'market' type presuppose a symbolic structure
which couples the difference between individual opinion and social
validity to the differentiation between means and ends: while the ends are
not subject to the process of exchange but left solely to the individual
will, the relations between **means** are the object of social control and

social interaction. The social validity and valuation of means differ systematically from the individual assessment of means or the particular utility of means with respect to a specific end in a specific situation. This is the focus of generalized media of interchange; they provide a standard which allows for a very precise description of the social validity of means quite regardless of individual preferences and particular situations. The vertical differentiation applied to differences between individuals in hierarchies is now transferred to the appraisal and assessment of means while the social relationship between actors in markets is regarded as basically symmetrical: superiority and inequality may be considered a contingent **result** but not a general **prerequisite** of competition in markets.

Markets too presuppose the differentiations generated by solidarity and hierarchy but these differentiations are ignored or reinterpreted in a radically different manner if actions are to be interconnected in a way which is typical for markets. The fact is well known from the classics of sociology: market interaction is rendered possible only if legal rights are accepted and if individuals who cannot be trusted are excluded from dealings. But these differentiations are applied only to demarcate the realm of markets and ignored if they come to structure the exchange process itself: as soon as the exchange relation is influenced by appeals to solidarity or claims on authority the characteristics of markets fade away.

Of central importance for the development of markets is the number of opportunities for interaction. Only to the extent that the number of mutually independent and **a priori** equivalent opportunities for interaction increases will the structure of markets evolve and provide situations favouring the use of generalized media of interchange. If contacts are confined to very few persons and if the relations between these persons are stable as well as affective and particularistic, then instrumental orientations to interaction and the application of generalized media for evaluating an action are regarded as inadequate for the continuation of their relationship. The successful application of media presupposes brief contacts, unrestricted opportunities and mutually independent offers of interaction. Otherwise the quantified comparison of means and goods rendered possible by the media code will not provide any significant

advantage over the direct exchange of goods on the one hand and
hierarchical control or distribution on the basis of solidarity on the
other.

Any privilege which restricts the number of individuals supplying or
demanding on a market not only contradicts the presupposition of equality
but also decreases the chance of **abstracting** from actual needs and given
offers with respect to future opportunities for exchange. These limits set
by monopolies to prospects of future opportunities obstruct the
application of the medium: confidence in the media code presupposes that
even if the goods offered to an individual do not appeal to him at the
time he will easily find an attractive offer at another time or another
place. If he knows however that the amount of exchangeable goods or the
number of individuals prepared to exchange is definitely limited, he will
be less inclined to trade useful goods for the symbolic representation of
opportunities for exchange. The successful application of media is based
on the confidence in unlimited opportunities for exchange. Observed
restrictions of opportunities, limitations of competition by monopolies or
personal solidarity impair the effects of generalized media and foster the
development of alternative codes of interaction.

VI. POLITICAL, SOCIALLY INTERGRATIVE AND SCIENTIFIC MARKETS

(a) If media codes are closely tied to market-type situations and
instrumental orientations to action, that is not to say that the
possibility of their being applied to non-economic problems is ruled out.
The functional - differenciation of problems of interaction - need
satisfaction via scarce goods and services, the collective establishment
of goals, social integration, and a binding interpretation of the world -
does not systematically lead to particular institutional solutions such as
the market, the state, stratification and modern science. Though the media
code presupposes that certain purposes are functionally differentiated,
and that an instrumental orientation to action and a market form of
situational structures exist, these conditions are in principle just as
likely to be present when political problems have to be solved as, for
their part, are vertical distinctions between social groups, moral and
normative orientations to action and hierarchical situational structures

when dealing with **economic** problems: **sovereign acquisition** based on legal
claims was, during long periods of societal development, the most
important path by which goods and services were transferred, and is still
familiar to the tax-paying citizen of today. The obligation to pay taxes
arises out of a typical situation which is distinct from the market: it is
based on an asymmetrical and lasting social relationship to which neither
the taxpayer nor representatives of the governmental tax-collecting
monopoly knows any workable alternative.[4]

Still older in terms of the history and the usual practice to this day
certain social situations is the distribution of goods according to a
principle of need or of equality. Mealtimes in the home, the sphere of
good-neighbourly assistance or everyday reciprocal agreements as to who
should use collective goods are well-known and commonplace examples of
such a "**communalistic**" method of dealing with economic problems. In these
cases the shades of the affective and personal, the lasting quality and
symmetry of the social relations involved coupled at the time with a
limitation of the alternatives are to the benefit of the inclusion code
and pose obstacles to the use of interchange media.

(b) The **solution to political problems**, on the other hand, is not
merely confined to enforcing **legitimate claims to authority.** Even if it is
true that state authority represents the most momentous solution in terms
of the history of development, and the most spectacular one in terms of
societal theory, non-hierarchical situations are nevertheless also
conceivable here, and indeed familiar to us all: ethnic ties, ties to
one's compatriots, party solidarity, membership in a family, solidarity
with one's own sex or ties between those of the same generation. A common
characteristic in all these situations of communal solidarity, apart from
equality among the comrades who share it, is the difficulty in changing
from one **association of solidarity** to another.[5]

Rather than discussing the code of inclusion in politics, we will
address a matter of more direct importance for our present purposes, the
question of the use made of general media of interchange in politics.
Parsons, as we know, analyses **political power** as such a medium of
interchange. In accordance with our hypothesis, we would only expect
political processes to be subject to control by media if politics itself
were functionally differentiated, if a market form of situational

structures and an instrumental orientation to action existed among those involved. In modern societies, a sphere devoted to things political is regarded as differentiated. We can most certainly impute an instrumental orientation to action to the majority of professional politicians, and the conditions cited for market-type situations appear to be fulfilled: **the number of votes** cast in political elections offers a **standard** with measurable characteristics and is hence formally equivalent to a currency; the political power acquired at an election is, in a way similar to money, relatively **unspecifically utilizable and can be circulative**, the exchange of political support for binding decisions or the promise thereof takes place under **competitive conditions**, and finally the **number of exchanging parties** available - voters and potential candidates alike - can tend towards **infinity.**

As slick an interpretation as this of political processes according to the market model risks being accused of superficiality. It takes no account of the fact that only a relatively small number of political decisions can be enforced in the short term on the strenght of majorities won during elections: **legitimate legal entitlements** place far stronger restrictions on the sphere of the politically electable than they do on that of the economically purchasable. The transfer-speed of the political commodity "binding decision-making" is limited by another factor too: While the buyer of a house or of stocks and shares can immediately resell them if he so desires, the "aquisition" of political decision-making rights in exchange for support as a general rule works on the basis that these rights are intended only for direct "consumption" by the person acquiring them. A group of politiciens who, let us say, have exchanged votes in an election for a particular office on the candidate's promise that a decision will be made in a certain way, can in principle dispose of this decision-making right to third parties in order to gain further political support in return. This procedure, quite usual in the everyday business of political compromise and the forging of coalitions, is not infrequently regarded by grass-roots voters as redolent of betrayal. The **moral barriers** confronting such a rapid turnover of political goods indicate that the voters, **i.e.** the "end-users" of those goods, do not always, or indeed even in the majority of cases, have an instrumental attitude towards political decisions. From the voters' perspective the situation of political action generally exhibits characteristics which can be categorized less as belonging to the market sphere than to the hierarchy or community types. If the politician deploys power or carries out authoritative functions, the legal code is applied and moral orientations are sought. If, on the other hand, he tries to establish legitimacy, then the orientation to action and the situational structure generally change in the direction of a community of solidarity. Though it is then the case that the voters and those they elect now meet on an assumed equal basis, the voters' attitude to the candidate is in all events shaded with personal and affective elements. This applies most of all when those casting the votes are not professional politicians.

One might raise the objection here that these limitations of market dynamics are of a historical and contingent nature, and that as political processes continue to be modernized they tend to be displaced to an ever greater degree. However there is a peculiarity about political goods which speaks against such an optimistic prognosis. Precisely that feature that identifies collectively-binding decisions and goals is the fact that they are **binding** over and above the exchange relationship, both temporally and socially. If a political decision-making right is continually traded, subjecting it to a variety of uses in quick succession, **this leads to a reduction in the value of the goods itself.** In contrast to material goods which do not suffer any marginal loss of use value if they are rapidly turned over in the market - a use value defined by an individual's relationship to the good **precisely** by ignoring society's evaluation - political goods do not have such a non-societal form of existence: It is only through social recognition that they become what they are. Hence, market processes in which political support is exchanged for decision-making rights run up against systematic barriers: To the extent that the situational structure of political action does take on a market form and facilitates the application of a media code, the goods being exchanged are threatened with spoilage.

(c) Analagous problems for the application of the media code occur when it comes to solving **the problem of communalization and integration**. In this case the inclusion code, that which belongs to a certain **mode of living**, has come to be the classic and appropriate solution whereas media and the legal code are seen as being problematic and crisis-prone. Though it is still quite possible to demonstrate the operations of hierarchically ordered **estate privileges** in everyday life in the modern civil service, industry and indeed the universities, the mere fact that they are demonstrable puts them under threat, being no suitable justification for them in the age of modernity. In developing a media code, interaction determinded by **influence** does not suffice as a point of departure, for it hardly has any structures of a market form in itself and, in particular, appears to lack any element equivalent to a currency. What can offer such a quantifiable standard of influence and communal attachment are **scales of prestige and status**, which do not sort and evaluate the material or social attributes of a mode of living according to criteria of money or of majority positions, but according to those of style, education and taste.

Such attributes of a common life-style can be transferred, deposited and accumulated. Even if the exchange metaphor does not appear to be altogether appropriate in this regard, style as an attribute is nevertheless the basis on which certain opportunities for communalization are allocated.

Although the rapid changes in fashions and styles occurring in modern societies do take on distinctly market-like-forms, the actual **characteristics of the goods being exchanged** (scarce opportunities for communalization in return for attributes of style) do nevertheless impose limitations on the development of "socially integrative" markets. As the value of a particular coin declines relative to the number produced, there is a similar fall in the prestige value of life-style attributes once they spread beyond a critical minimum which was necessary to generate discussion over differences in prestige value in the first place. This devaluation generates constant pressure for new variants of life-style which, only through a relatively restricted initial distribution, can secure a high prestige value. If the rapid succession of styles and fashions is interpreted according to the monetary model, one would have to term this a dramatic in-built **inflationary tendency** leading to the continual launching of new currencies and to **rapid decay in the media code**. This inflationary tendency seems hardly to be stoppable: In contrast to markets in goods and services where the money supply is subject to control by a central bank or is naturally limited by the scarcity of precious metals, any restrictions on the choice of those style attributes deemed to create new communities can only be enforced in a roundabout way: either by way of the market price demanded for the attribute of style, or by way of legal rationing systems which obviously have little change of being accepted in modern societies. The **free will** upon which any communal association purged of all economic, political or ideological obligations is based thus prevents the emergence of a prestige order binding for all members of a society. Instead, there are invariably at any one time in modern societies a whole series of "communal currencies" in circulation, which within a short time find acceptance, broaden out and then lose their value as a result. Attributes of style differ from coins in that they cannot be circulated, but are passed on by **replication** and loose their scarcity as a result of the transfer process.

What is more, it can be assumed that rapidly changing fashions for all their prominence, only convey a limited proportion of the relations within communal association. What are likely to be more significant for communal relationships between two actors are those differences in taste, style and **education** which led Bourdieu to speak of "educational capital" but which, unlike monetary forms of capital, cannot be readily transferred, can only be acquired through painstaking educational processes and usually remain with a person for a lifetime (Bourdieu, 1982). In the case of education, the slow pace of reproduction thus prevents a market form of exchange processes from coming into play.

Finally, there is a tendency inherent within community-related interactions for instrumental- and functionally- specific orientations to be given up relatively swiftly in favour of affectively-shaded and functionally-diffuse attitudes. The situational structure then "slides" systematically, as it were, from the market-like level to the hierarchical level where differentiation is based on estate, or to the level of **life-world** personal relationships. Economic markets have a tendency towards abstracting real value bases. In the case of the market for communalization opportunities, however, the opposite evidently applies: they require promises of communal integration to be continually honoured in concrete life-world relationships.

Reasons for this can be seen, apart from in the inflationary tendencies and transfer problems mentioned, in the function of communal association **per se:** if a group is formed without any **a priori** consensus as to the goals of collective action or without any expectation of individual gain, an injection of confidence which any interaction requires is possible. This also determines the limits of necessary group attachments: they coincide with the scope of possible interactive contacts. The inexorable tendency for community relations to be expanded is **not**, then, in the interests of the actors concerned. This is true not only because this expansion is accompanied by a growth in the sphere where duties of solidarity prevail, but also because multi-faceted community bonds encourage the emergence of **solidarity conflicts** which place a question mark over the stock of communality as such. The media code, which can become truly effective precisely when interactive contacts are unrestricted, only gains a "superficial" significance in view of these tendencies for communalization processes to be limited: It facilitates a sense of community in brief encounters between strangers, but becomes superfluous and tends to be more of a nuisance as the communal bond between actors grows and is built upon.

(d) A situation which appears more favourable towards the development of market processes is that under which interpretations of the world become binding in modern societies, namely **science**, as guided by **methodical arguments**. The latter seek validity or acceptance as criticisms or justifications, **i.e.** as evaluations of interpretations of the world. If an interpretational or explanatory proposal is not accepted, or is accepted only to a limited extent by a community of scientists, its level of acceptability may in principle be raised if additional methodical arguments are supplied. However the opposite applies only with reservations: If methodical arguments are proposed which **oppose** an explanatory or interpretational proposal, this only weakens the likelihood of acceptance if an alternative is available which can be more elegantly and satisfactorily proven. In this respect there is evidently a parallel between scientific markets and those for goods and services: it is only when some means of substitution is available that price increases lead to a reduction in demand. An institution to protect critical and/or substantiating claims also appears to exist, namely the authority of scientific **paradigms**. The institutionalization of paradigms itself is clearly just as little a product of market-like conditions as is the authority of central banks or constitutions. Agreement with the claim made

by a paradigm represents the foundation of methodical criticism in just
the same way as recognition of the constitution is the precondition for
political elections, or as central-bank authority stabilizes currency
values. The pledge to maintain an impersonal approach and a functionally-
specific rationality in scientific action is another factor which
facilitates the application of a media code. Yet another is the
opportunity for communication associated with a methodical argumentation
which in principle ranges across the entire breadth of a paradigm.

An objection might be raised on the grounds that the scenario of
market-like critical processes portrayed above is a methodological
idealization which only accords with the real-life influences controlling
science - career interests, politics and instructional authority -, in
marginal cases. However, such reference to factual behaviour which
deviates from norms does not affect any hypothesis relating to the
validity claim of norms - in this case the media code of "methodical
arguments". As far as the validity of a monetary code under market
conditions is concerned, the development of **factual** disequilibria (not
those due to legal form) in individual exchange relations or an
orientation by individual market participants to long-standing business
relationships is not sufficient to impair that validity. A code of
interaction does not become null and void until it can be shown that,
under certain conditions with selective effects, it leads to systematic
communication crises. However this is not the case either for the monetary
code in economic markets or for the methodical argumentation code in
scientific discourse.

The limitation on the development of market-like situations in modern
science is not so much the behaviour by academics deviating from the path
of methodological virtue, as it is the absence of any binding **quantifiable
standard**. For although methodical arguments can be accumulated and, for
that matter, can also be brought into an order of priority, they can
hardly be precisely measured and compared. The impressive attempt mounted
by the programme of inductive logic to develop such a standard of
empirical confirmation ended, as we know, in failure (Lakatos, 1968, pp.
315-417). This absence of a readily applicable quantifiable standard and
doubts within the scientific field regarding the ideal of an empirically
substantiated unitary science bring with them the consequence that

scientific critical processes can develop a media-like code only on the basis of a narrowly cast paradigmatic consensus: common formulations of questions as a reference for methodical arguments, the quality of which can be assessed according to the precision and scope of the relevant data. However, broad paradigmatic consensus when it happens is frequently confined to a relatively small group of scientists, and their own personal relations pose another problem for the application of a media code. On the other hand, broad-ranging discussions transcending the boundaries of "normal science" which also receive attention outside the academic world rarely make any use of a standardized code. Though market-like conditions do appear to exist in this instance, the processes of communication tend rather more to follow the model of the "bazaar": Although the acceptance of interpretational proposals is not dependent on legitimate claims or inclusions, it nevertheless only runs its course via circumstantial and long-winded persuasive efforts which stand in the way of the rapid transfer expected of a market.

There is another important aspect in which methodical arguments differ from media such as money or election votes, namely that they are not **scarce** in the sense that the use of an argument to substantiate an explanatory or interpretational proposal does not rule out that same argument being used again, or simultaneously, in providing a basis for other proposals elsewhere. While the very thing giving money its value is that the volume in circulation cannot be reproduced at will and thus cannot both be used to acquire some goods and yet be retained, the opposite is the case for methodical arguments: They tend to become all the **stronger** the **more** they are used by other scientists and hence achieve recognition. Criticism and processes of substantiation work according to the rule that while methodical arguments are indeed scarce, they do not need to be kept scarce by social control. This indicates problems in differentiating between the real value represented by "interpretation of the world" and the opportunities for communication determined by the medium of "methodical argumentation". In contrast to political power and communal association, where any such separation of a non-societal form of existence (the real value) from the exchange form arising in the market process would be a fundamental impossibility, it actually appears quite conceivable to differentiate between rational validity and social commitment in the case of methodical arguments. In this case, however, the situational conditions surrounding the media code's application in the field of modern science are more favourable than they are with regard to communalization processes or politics; what does appear to be lacking is simply a standard of methodical substantiation.

VII. CONCLUDING REMARKS: MEDIA BETWEEN INFLATION AND DEFLATION

If media of interchange are regarded as one of a number of codes which are simultaneously possible in controlling interactions, the question arises as to which empirical conditions will prove favourable to the application of the media code. The answer to the question that has been given here assumes that media of interchange are successful in abridging processes of communication as far as the situation of interaction possesses the formal aspects of a market, though it should be noted that such market situations are not restricted **a priori** to solving economic problems. In attempting to analyse the possibilities for the development of political, socially integrative and cultural markets, it soon became evident to us that the potential for such markets is actually restricted, and that there are therefore empirical limits to the application of interchange media in politics, communal association and culture in the light of the institutions currently known and accessible. These limits can be summarized under the following three headings:

(1) The absence of any **measurable standardization** and restriction of the **number of interacting parties** who can be addressed in a particular situation prevents any rapid and precise comparison of the products offered for exchange or strips them of any advantage: the **abridgement of linguistic-bargaining processes** which media make possible has no effect in such cases. If only a limited number of exchanging parties is available, the **confidence in abstract exchange opportunities** develops only with difficulty. The ability of media to **rapidly** mediate between and process **far-reaching** interactive contacts, **i.e.** to accelerate the rate of reproduction, does not provide any selective advantage under these circumstances.

(2) If the transfer of knowledge and norms conveyed by the media does not occur as the **circulation** of rights and claims, but as the replication

of rules and knowledge, then one automatic characteristic of market exchange no longer exists: Exchange always presupposes that the good involved is actually passed from one owner to another and does not become available to both exchanging parties at once. Money and election votes do fulfil this condition. Style attributes and methodical substantiations however do not. The ability of media to circulate is closely linked to the **institutionalization of controls of coverage.** Democratic constitutions and central banks exert a controlling influence over the relationship between the claims and real values mediated by the media which applies to **all** interactions where a given currency or electoral procedure is used. An equivalent control over scientific discourse by paradigms relies on much less stable foundations, and one over style attributes is almost completely absent. This weakness in protective controls leads to **inflationary developments** which cast doubt upon the effectiveness of media as reproductive codes. The expenditure of media - of style attributes, methodical substantiations, electoral votes or money - needed to swap certain real values can increase to such an extent that media are no longer an "abridgement" of interactive processes, but actually complicate them.

(3) A crisis front at the other end of the spectrum where media are selectively successful comprises the **deflationary limit** of media dynamics. Media deflation arises because of the **effectiveness of other reproductive codes** which draw the transfer of claims and attachment to rules away from guidance by the media. In given situations, the code of inclusion and the legal code operate more successfully than the media code in situations involving communities of solidarity, or hierarchies. The question as to why media dynamics are confronted with such deflationary barriers more frequently, as empirical experience would have it, in politics and social integration than in economics or the sciences thus becomes extended to a search for the factors working against change towards market-like structures in communities of solidarity, and in hierarchies. An answer suggests itself if we point up the fact that the constitutive conditions for collective decisions and social interpretation are, in both cases, **social** in nature. Whereas the utility of economic goods and the verity of scientific statements - at least in our conception of them - can retain their value to an individual even if no other individual interacts with

him/her, this is not the case for socially binding decisions or for communal association. They need the attachment to their real value basis in communities of solidarity and in hierarchies; the price of transforming this in markets is the destruction of the real values.

Habermas, then, appears to be right on **this** point: the nature of interactive problems posing themselves in particular situations contains critical conditions for the development of media. The precondition for analyzing these conditions, however, is not a categorial-analytical model, but an explanatory programme using selection theory: the latter is directed towards empirical situational conditions and problems rather than towards any growth in rationality established **a priori**.

NOTES

1. For example E. O. Wilson and others in biology, D. Campbell in psychology, K. Boulding, F. Hayek and J. Röpke in the realm of economics, K. R. Popper and S. Toulmin in the philosophy of science.

2. Two different contexts may be distinguished. In the first case media are used as functionally-specific symbols for control of resources although the interaction between the persons applying the code is controlled by this medium. Officials administering money according to legal rules are an illustration of this case. Money is subject to actions but does not control these actions. If however officials can be motivated by money to make a particular decision, this money controls and coordinates their actions. It is mainly this function of control that interest Parsons and Habermas.

3. If monetary recompensations take the place of matrimonial relationships of solidarity, problems of communications or even a dissolution of the matrimonial relationship will be the consequence.

4. Nevertheless departures from existing practice in the direction of market-form situations are conceivable: **Steuerpachtsysteme**, from the government point of view, or establishing permanent residence under a jurisdiction with favourable tax conditions from the tax-payer's point of view, offer restricted opportunities to exercise choice in the market.

5. Precisely because the effectiveness of the code of inclusion can also be assumed to be assured in the sphere or organized and institutionally differentiated politics, it does not receive moral support in this area: if anything, particularistic attachments here are regarded as dubious on the grounds of possible corruption.

RICHARD PIEPER

THE SELF AS A PARASITE
A SOCIOLOGICAL CRITICISM OF POPPER'S THEORY OF EVOLUTION

I. INTRODUCTION

Parasites are creatures who survive at the expense of others. An
evolutionary transition to symbiosis is not unusual and the dividing line
is blurred. In any case, a distinction between two different species
existing relatively independently of each other is assumed, as opposed to
the internal relations of an organism. Thus, we may have a potentially
fruitful model for the kind of interactionism and dualism represented by
Popper. In his book, coauthored with Eccles, **The Self and Its Brain**,
Popper in fact attributes characteristics of parasites to the self. Popper
himself does not use the term, and presumably he would prefer to describe
the relationship as symbiosis. But certainly we may assume that biological
analogies would meet with his consent, since he explicitly claims a
"biological approach".

The self resembles a parasite in two ways in Poppers's theory. First,
the autonomous, creative self appears as a "ghost in the (biological)
machine" [1], and the brain as an instrument of survival. This dualism
suggests that there are virtually two different species. No doubt the
sympathies of the liberal critical rationalist Popper are with the self,
which uses the organism for its higher purposes even, we may add, at the
expense of the health (smoking) or life (martyrdom) of the organism.
Second, we recall a favorite formulation of Popper's on the significance
of theories in the struggle for survival: We can let our theories die in
our stead. We live parasitically from the success of our theories, which
themselves fight a quasi-biological struggle for survival in the evolution
of knowledge.

M. Schmid and F. M. Wuketits (eds.), Evolutionary Theory in Social Science, 195–224.

But are these biological analogies really tenable? Can we actually divide body and mind along such dualistic lines? Is an evolutionary theory of the body-mind relationship not necessarily monistic? And what about the possibility that theories and ideologies actually have us fight their struggle for survival for them? There are martyrs, revolutions and religious wars. Paradigms may "wait" in the process of science until their opponents among scientists die. Ideas can in fact be "resurrected" while the scientists themselves cannot. Are ideas or "selfish memes" (Popper's cultural objects) not the actual parasites competing with the "selfish genes", as the sociobiologist Dawkins (1981, p. 143) has it? Such a perspective corresponds to Popper's "epistemology without a knowing subject", or his "theory of the objective mind" (1972).

Hence there is an unresolved tension in Popper's theory. In the context of his arguments for the dualism of brain and mind, he stresses precisely the autonomous role of the individual which he attempts to minimize in the context of his theory of knowledge. Especially the biological analogies create the problem that the creativity of the self, so highly valued by Popper, threatens to degenerate into a series of random mutations in an objective selection process. The autonomy of the self appears to be self-deception. The problem comes down to the question of what exactly the nature of the **interaction** is supposed to be between body and self on the one hand, and the self and sociocultural objects (knowledge) on the other. The problem reappears on a more general level in the relation between an "upward" emergence and a "downward" causation in the hierarchical structure of Popper's three worlds.

My present concern is not to go into details of the philosophical discussion of the body-mind problem. Popper himself refers only cursorily to the extensive literature and thus makes an evaluation of his is hints and conjectures in the light of this discussion rather difficult.[2] Let us note instead that in his basic intentions Popper actually will find hardly any opponents among adherents of the predominant sociological approach, the theory of action and action systems, although they may disapprove of the fact that he ignores sociology altogether. Inasmuch as Popper's interactionism and dualism are designed to restate the "relative autonomy" (Durkheim) of social phenomena and reject reductionism, most sociologists would agree with his position. The problem arises with more precise concepts and models of interaction between Popper's three worlds, with the place of sociology in his scheme and with the question as to what extent Popper actually departs from traditional monistic **or** dualistic positions, which he takes to be incompatible with his notion of an interactive dualism. My argument is that, at least from a sociological perspective, Popper perpetuates in many respects a dualistic tradition incompatible with his better intentions.

I will proceed as follows. First, we have to clear up some confusing arithmetic in his presentation, especially if earlier formulations are included. He argues **pluralistically** for **dualism** in the context of a **Three-**Worlds Theory that contains at least **seven** so-called cosmological stages. Popper switches here from the perspective of evolutionary theory to epistemology in a way which leads to confusion.

Second, Popper argues against reductionism and for emergentistic indeterminism. In this frame of reference Popper develops one of the most important and, at the same time, least recognized features of his theory. Popper proposes a kind of ecological cosmology which, in central aspects, owes more to his interpretation of indeterminism in modern physical theory than to the theory of evolution. Here we find remarkable parallels to Durkheim, who also accepted dualism and an evolutionary perspective, but rejected the biologistic versions of the theory of evolution he found in Spencer and Schäffle. Durkheim turned rather to physiology and chemistry for analogies of creative emergence in evolution. Unfortunately, Popper hardly mentions modern theories of emergent evolution or self-organization in physics, chemistry or biology which could help to clarify his concept of interaction. A discussion of these theories would exceed the limits of this article, but we will have to analyse the strains in Popper's dualism as they arise from his own theory of emergentistic indeterminism.

Third and central to a sociological appraisal is an analysis of the precarious of the self as a wanderer among Popper's worlds. His "biological approach" will prove to be rather helpful in his criticism of neurophysiological "materialistic" theories. But his claim to a "biological approach" remains dangerously close to a reductionistic sociobiology. Following Durkheim, I will consider dualistic interactionism in a general evolutionary perspective rather than within the narrow limits of a **biological** approach.

II. DUALISM, TRIALISM OR PLURALISM ?

In his Three-Worlds Theory (TWT) Popper distinguishes between a world of physical processes (W1), a world of subjective, mental or psychological processes (W2) and a world of objective ideas, theories and other cultural objects (W3). The three worlds interact, although W1 and W3 interact only by means of W2, i.e. theories transform our physical world only by way of the mental activities of man. Social phenomena such as language, culture or institutions are considered elements of W3 without further explanation. The interaction is understood to be a causal, indeterministic relationship; the processes within these open and only relatively autonomous worlds are characterized as causal also.

That is, W3 appears to be an exception to these causal relations, since Popper speaks of logical internal relations only in W3, and thus mystifies the relation of sociology to W2 and W3. In the sociology of knowledge the problem of the relation of the social individual to stocks of knowledge has some tradition as the question of **Seinsverbundenheit** (Mannheim). Although Popper's vigorous rejection of the sociology of knowledge as "subjectivism" is well founded in the context of his theory of scientific knowledge, it tends to push social phenomena into the subjective W2 and, in any case, does not help to clarify the position of sociology, especially the sociology of knowledge, in his TWT.

To understand Popper's theory we have to consider tree problems which he tries to solve within this framework. His main problem is to buttress his objectivist and realist position in epistemology. Some commentators and critics even see his TWT as nothing but a quasi-ontological foundation for his logic of scientific discovery.[3] The evolutionary frame is expected to explain why our intrinsically fallible knowledge turn out to be so surprisingly successful and capable of growth. Darwinism furnishes Popper with a more or less metaphysical research program which, on the one hand, places his objective W3 in a historical process opposed to Plato, and in this respect Popper himself acknowledges a closeness to Hegel (1972, p. 125). On the other hand, his cosmology or natural philosophy is intended to exclude any subjectivism or idealism. The special character of W3 in being close to W1 and W2 within a common evolutionary background but separated, particularly from W2, for epistemological reasons is explained

by its function in this framework. However, as I understand Popper, he is
not especially interested in more or less metaphysical foundations, but
tries to develop his arguments primarily within a scientific theory of
evolution. For this reason, the precarious scientific character of
Darwin's theory, **i.e.** its questionable falsifiablility, made Popper reject
it at first, and only in his later works did he take a "darwinistic
turn".[4]

In "Of Clouds and Clocks" (1972, pp. 206-255) Popper distinguishes two
additional problems within this evolutionary framework: Descartes' problem
and Compton's problem, as he calls them. The first pertains to the
relationship of body and mind, the second to the possibility of control of
physical or physiological processes by ideas, norms or theories. The first
problem appears in our purposeful actions, the second in the realization
of quasi-platonic ideas or plans in material objects. Popper is right in
stressing that Compton's problem is more fundamental, since it must be
solved independently of available psychological theories and poses a
problem for any of them. He approaches the problem by proposing an
indeterministic theory of the universe as a hierarchy of "plastic"
controls from a subatomic up to a sociocultural level. This model of
hierarchical, open and interacting systems he considers a solution to
Descartes' problem. Only in later, again more epistemological articles
(1972), does Popper reduce this hierarchy to a tree-world scheme.

What is happening? Apparently Popper is unsatisfied with his model's
capacity to grasp the specific role of the self or ego in Descartes'
problem. In addition, he is still laboring with Compton's problem and so,
for want of a satisfactory solution, he projects each of the three aspects
of these problems into a "world" and places W2 between W1 and W3.
Obviously, Popper discourses throughout without much hesitation about
relatively independent worlds whenever some important, yet unresolved
problem somehow subdivides the realm of scientific inquiry (1972, pp. 289-
295). The original model of a control hierarchy arises in the context of
an argument against physical determinism. Being by conviction an
indeterminist in the interpretation of modern physical theories, he
intends to prove physical determinism untenable. Deterministic systems are
necessarily "closed", he claims, and are therefore incompatible with any
reasonable concept of interaction between the physical world and the world

of ideas, **i.e.** with a conceivable solution to Compton's problem. Thus Popper considers indeterminism to be a necessary condition for any theory of "open", interacting systems. To be sure, indeterminism is not enough - as Popper maintains in a well-known phrase. Rational arguments and decisions especially, are not random events. But, in any case, indeterminism is defended with reference to the interaction of causal relationships, that is, a basic unity of different realms of science.

Popper's concern is quite different when proposing a tripartite universe; here he confronts the problem of reductionism and monism. His argument consists basically of a defense of a critical, pluralistic methodology and a robust realism, **i.e.**, a claim to reality is acknowledged for everything we experience as a real object in our everydaylife and which can be demonstrated to interact causally with prima facie realities (1977, p. 9). This includes a real existence of selves (p. 101), whereby Popper does not distinguish in any relevant sense between a "solipsistic" ego and other selves or minds. Now any attempts at a "premature" reductionism violate his critical pluralism, since in Popper's view such reductionism **either** rests on some metaphysical monism unsupported by the obvious heterogeneity of the subject matter of science, **or** pursues some utopian monistic research program which overlooks the fundamental fallibility and incompleteness of our knowledge. These limits to our knowledge restrict us to a pluralistic program and a critical evaluation of alternative theories. Successful reductions happen rarely if ever. Moreover, complete reductions presuppose a complete description and prediction within both the reducing and reduced theory which is impossible in principle (1974b). Finally, Popper's realism leads to the assumption that whatever our most successful theories take to be real does exist and will interact with other things in the hierarchical structure of the universe. "Reduction in principle" in Popper's view is incompatible with a "downward causation" **i.e.**, with causal interaction in the hierarchy going in **both** directions (1977, p. 19).

Popper combines here arguments for pluralism and realism in a way which may refute traditional reductionistic and monistic materialism, but these arguments are not sufficient to support his stronger thesis of a hierarchical order of worlds. The epistemological arguments for critical, fallibilistic and pluralistic realism work against reductionism, but not

for his proposed **structure** in the universe of irreducible domains.
Besides, no contemporary reductionist will assume a **complete** reducibility
as a realistic goal. He will only maintain the fruitfulness of a
reductionistic research program. No behavioristic or physicalistic
reductionist will be overly impressed by Popper's arguments, since they do
not counter the criticism of psychological terms such as "mind", "self",
"self-identity", "consciousness" or "mental maps" as being imprecise or
superfluous. Popper himself makes inflationary use of mentalistic concepts
and becomes vulnerable in his interactionist position, since it rests on
the availability of a respectable theory of subjective mental processes to
constitute W2. Popper himself does not present or discuss such a theory,
although one may take Eccles' contribution to fill the gap.

Darwin's theory of evolution solves the problem of hierarchical
structure to some extent for Popper (as for reductionists). The mechanics
of mutation and selection produce new objects and their reality is
confirmed in the struggle for survival. Popper employs evolutionary
arguments against any epiphenomenalism, since evolution does not produce
useless phenomena having no effective, causal functions.[5] Evolution may
explain the development of hierarchical structures, but it does not by
itself support the assumption of "ontologically distinct worlds" (1972,
p. 154). Further more, the demonstration of an evolutionary logic of trial
and error **within** different domains, e. g., the development of biological
organisms and the history of scientific theories, does not show an
evolutionary relationship **between** such domains. Thus, Popper still has to
explain just why he does **not** conceive his hierarchical structure of levels
or worlds monistically and why he proposes **exactly three** worlds.

In discussing specific reductions Popper apparently treats any
evolutionary level as a "world" which may or may not be, in fact,
reducible to a lower level. He mentions seven such levels or "cosmic
evolutionary stages" and maps them onto his three worlds (1977, p. 16).
Yet on another occasion, he explicitly admits the possibility of
distinguishing further worlds (1974b). Apparently there are first-class
worlds and second-class worlds in Popper's scheme, although he himself
does not offer any systematic distinction. The TWT pertains to the first-
class worlds; his pluralistic arguments against reductionism allow for any
secondary subdivision our ignorance or the progress of science may call

for. Thus the emergence or creativity in the evolutionary process produces many levels and boundaries, but those boundaries between Popper's three worlds are apparently of a qualitatively different character from the others.

Following a distinction by Wimsatt (1976), we may interpret Popper to assume a difference between an emergence within a specific world and between two worlds. **Within** worlds we can ground our hopes reductions in the development of more encompassing new theories. The crucial concept would be the **between-worlds-emergence** or the **relative autonomy** of worlds, which is not (only) attributable to contingent, historical defects in our theories, but (also) to more fundamental barriers to our knowledge of the structure of reality. This "cutting problem", as we may call it, or the question of where and on the basis of which criteria Popper introduces irreducible, emergent levels justifying assumptions of first-class worlds with relative autonomy, will guide the further analysis of his dualism and interactionism. We need a concept of emergence which is not only the projection of our ignorance onto the world. Popper, in fact, offers a concept which could well serve this purpose, namely, the concept of propensity. This potential function of the concept of propensity is not recognized, as far as I can tell, in the reception of his TWT. We will try to illustrate it in the context of Descartes' problem, which Popper himself chooses in his discussion of the self and its brain.[6]

To sum up, the confusing arithmetic results from the fact that Popper argues **pluralistically** against reductionism, defends **dualism** primarily in its anti-reductionist qualities, and introduces a **third** world to harbor objective knowledge (and other cultural objects) at a safe distance from "subjectivism". His view of the problem may be summarized in the following way: The structure of domains and their relationships may be conceptualized differently, but in any case they must have emerged in an evolutionary process from each other, **if** we accept Darwin's theory as a cosmological research program. This process must be conceived to be fundamentally indeterministic, **if** we do not question modern microphysics. In this process there must be an emergence of "real novelty", **if** we accept the irreducibility of W2 and W3, **i.e.** matter must be assumed "to transcend itself". Finally, no world in an evolutionary theory can be only an epiphenomenon of some other world, and thus their relationship must be one of causal interaction.

III. DESCARTES' PROBLEM

But what kind of paradigm or model do we have at our disposal to characterize this interaction between worlds or emergent levels? Such a model must be applicable, for instance, to the relationship between neurophysiological and psychological phenomena.

Disregarding epiphenomenalism and metaphysical monism, as Popper does, we are left with physical reductionism, panpsychism or dualistic interactionism as alternative paradigms. Since reductionism is inacceptable to Popper and panpsychism appears to be highly implausible on the lower levels of the hierarchy, interactionism remains with us as "default view" (Economos, 1978), if Popper can present a plausible model for the interaction. Since he argues mainly against philosophers of the past, we will consider briefly some contemporary theories on the bodymind relationship to demonstrate a certain legitimacy in his restriction of the discussion to these alternatives. The controversy between Bunge, MacKay, Sperry and Eccles is chosen for this purpose because it refers explicitly to Popper's position.

Bunge (1977) argues for an "emergentistic materialism" and postulates an explanation in principle of emergent qualities through relationships between lower level elements. Popper rejects such an "explanation in principle" (1977, p. 19), since it logically rests on a complete "description in principle" and a "prediction in principle". Moreover, an explanation in principle "from below" implies that there is only causality "from below" (1977, pp. 19-21). But Popper insists on interaction and therefore on the existence of "downward causation", a concept he adopts from Campbell (1974) and Sperry (1976) and which states the causal effectiveness of upper levels for lower level processes. Gravity serves as a physical example (1977, p. 20), but Popper applies this concept also to the effects of consciousness on neural processes, of the environment on the organism, and of theories on our actions.

Restriction to a causality "from below" induces a strategy of ascribing properties of systems to their elements as dispositions. But as Sperry remarks, "The reductionist approach that would always explain the whole in terms of the parts leads to an infinite regress in which eventually everything is held to be explainable in terms of essentially nothing"

(1976, p. 67). Now Popper himself acknowledges that regresses may unavoidable. What he finds questionable is that reductionism arbitrarily stops this regress of differentiating wholes or the corresponding progress of aggregating wholes. If there are no explanatory hypotheses, it is only an ad-hoc strategy to ascribe, say, to neurons the disposition to show consciousness in some appropriate "assembly" (Bunge).[7] First, it may be theoretically more fruitful to ascribe the disposition to the system or even to an encompassing system-environment context. Second, as Sperry indicates, there is **prima facie** no reason to consider physiological states as **objects** and psychological phenomena (only) as their **properties**. The postulate of "explanation in principle" apparently is intended to guarantee a materialistic monism of "objects".

If the cutting problem is solved by Bunge in a traditional, materialistic fashion - horizontally, as it were - between the domains of natural and social sciences, MacKay (1978) chooses a kind of vertical cut, **i.e.** cybernetically between forces or energies and form or **information**. According to MacKay, mental processes are a special case of the fact that psychical processes **also** have an informational aspect. There is no interaction between worlds, but only two different compatible ways to describe and analyze processes. The relationship betwenn physical and physical processes is one of carriership or embodiment, not one of interaction between different "objects" (MacKay, 1978, pp. 602-603).

This position is typical for many approaches employing the analogy of mind/program/software and brain/computer/hardware. But the epistemological possibility of alternative descriptions is irrelevant to the question of genetical or evolutionary relationships between different phenomena, as Bunge (1979a) objects to such "double-aspect" theories. The hierarchy of physiological and psychological levels is not a hierarchy of analytical abstractions. Popper does not discuss these approaches in cognitive psychology and artificial intelligence, although he explicitly rejects computer analogies. This is unfortunate, since Popper's rather vague alternatives are presented in this discussion in considerably more precision and detail.[8]

However, Popper's rejection involves a central problem of these analogies. Computers are construed by man to realize specific information processes allowing for a clear distinction between machine and program, between organ und function. A "panpsychic" character of the machine is guaranteed by the preceding logical differentiation of cognitive capacities and an allocation of elementary functions to elementary machine parts. Thus we may say that a powerful psychical "downward causation" is effective in the construction process. Computers are clocks and not clouds, as Popper remarks, because only strictly deterministic, closed structures can provide a flawless implementation of programs.[9]

Although Popper rejects computer analogies, we may illustrate his model of the mind-brain relationship with the interdependence of a programmer of average intelligence to a computer with technical defects (indeterminism!). The kind of program that is realized in this interaction depends on the specific errors made by both programmer and computer. Two aspects of this analogy are relevant: First, the respective theories of errors are substantially different; thus we have a dualism even of the "irrational" aspects of both systems which supports the analogy. This is a recognized fact in computer science and requires **empirical** research into the error behavior of different physical realizations (Nelson, 1975). Second, there is an important difference between computer and brain that is not covered by the analogy, since we are to assume that the brain is influenced in its physiological structure by its forms of usage. The brain learns not only in the sense of modifying its "software", but also by changes in "hardware" structure. There is no clear-cut distinction. The causal interaction between levels implies, as Popper points out, that to some extent each self during its development produces its very own brain or "computer" (1977, p. 473).

The problem with computer analogies is that they apparently demonstrate how completely mindless, material silicon chips may realize mental capacities by appropriate combinations. This seems to make some more or less reductionistic monism more plausible. What is missing in the analogy is a theory about the **natural** mechanisms of combination. Reductionists like to refer to future progress in biological theories of evolution. But at present these theories are completely incapable of explaining the evolution of mind. This holds also for recent theories of autopoiesis, even though they pose the problem in ways which are very close to Popper's intentions.[10] Moreover, reductionists should recognize in their references to computer analogies that there is a strong tendency toward dualistic

positions in computeroriented cognitive psychology. Typical cases are the
Cartesian dualism of Chomsky (1980a; 1980b) in linguistics and the
computational-representational view of Pylyshyn (1980) in cognitive
psychology. Like Popper, Chomsky, for instance, adopts a realistic
interpretation of mental functions, even if it may turn out to be utopian
to identify the corresponding and interacting neurophysiological states,
as Putnam (1975) argues. Therefore, Compton's problem should prove to have
no solution in the sense of an explicit theory of the interaction of
language and brain, and the dualistic cut between worlds would be
explained as a practical epistemological barrier. But these arguments
still acknowledge a "downward causation", as Sperry points out (1978, p.
366).

Our present knowledge of neurophysiological processes in the brain does
not warrant a simple identification of mental capacities with functions of
neural networks in close analogy to the relationships in a computer. This
is a common point of departure for both Sperry and Eccles. The unsolved
problem of localizing mental functions in the brain suggests rather that
there are still unrecognized, higher or emergent levels of organization of
brain processes with characteristics of force fields. Sperry interprets
such emergent levels realistically and assumes a causal interaction
between levels, as in the relationship between physical and chemical
entities. In this regard, Sperry recognizes close parallels between his
position and Popper's hierarchy of plastic controls.[11]

Sperry at least suggests how we may conceptualize processes of
emergence in a monistic framework. He explains holistic properties with
reference to the space-time configuration of system elements, which he
assumes to be an influential factor in the constitution of the system.
Elements are parts of encompassing systems, environments or fields and
cannot arbitrarily be abstracted from their ecological context - neither
in practice nor theoretically.

Neither Popper nor Sperry follows Eccles in his speculations on a kind
of wireless radio transmission relationship between mind and brain which
comes dangerously close to parapsychology.[12] But Popper recognizes that a
fundamental problem arises regarding the relationship between natural laws
on different levels, and the possibilities for the emergence of new laws.
At this point he introduces his propensity interpretation of probability,

an interesting and strategic move unacknowledged by his critics and commentators. This justifies an extended quotation:

"The usual materialist and physicalist view is that all the possibilities which have realized themselves in the course of time and of evolution must have been, potentially, preformed, or pre-established, from the beginning. This is either a triviality, expressed in a dangerously misleading way, or a mistake. It is trivial that nothing can happen unless permitted by the laws of nature and by the preceding state; though it would be misleading to suggest that we can always know what is excluded in this way. But if it is suggested that the future is and always was foreseeable, at least in principle, then this is a mistake, for all we know, and for all that we can learn from evolution. Evolution has produced much that was not foreseeable, at least not for human knowledge." (1977, p. 15)

"We can admit that the world does not change insofar as certain universal laws remain invariant. But there are other important and interesting lawlike aspects - especially probabilistic propensities - that do change, depending upon the changing situation. (...) There can be invariant laws and emergence; for the system of invariant laws is not sufficiently complete and restrictive to prevent the emergence of new lawlike properties." (1977, p. 25)

"But while the invariances may continue to hold for elementary physical entities (atoms, inanimate structures) sufficiently distant from the newly emerged structures, new types of events may become the rules within the fields of the newly emerged structures; for with these emerge new propensities, and new probabilistic explanations. (...)

An interesting objection to this argument has been raised by Jeremy Shearmur: even if we admit propensities, we do not escape the idea of preformation - we just have several preformationist possibilities instead of one. My reply is that we may have an infinity of open possibilities, and this means giving up preformationism; and this infinity of possible propensities may still rule out infinitely many logical possibilities. Propensities may rule out possibilities: in this consists their lawlike character.

I suggested something like this many years ago, in an attempt to explain the world view of the propensity interpretation of probability in my still unpublished Postscript. The infinity of the inherent possibilities or propensities is important, since a probabilistic doctrine of preformation does not otherwise differ sufficiently from a deterministic doctrine of preformation." (1977, p. 31)

As Popper points out in a different context, propensities or objective probabilities are not properties of things, nor their dispositions or intrinsic forces in the sense of Aristotelian potentials (1978b, p. 259). Propensities have the characteristics of force fields; they pertain to the situation as a whole. We should realize that Popper at this point continues an old battle against "essences" and "substances" employing a (meta-) physic of events, states and processes with reference to the dissolution of material things in modern physics (1977, p. 7).

In a biological interpretation, propensities pertain to the ecological relationships between system and environment, not to properties of organisms. Neither are they to be explained reductionistically as effects of local, elementary interdependencies by a causality "from below", nor should we interpret them subjectivistically as effects of ignorance, i. e., as expressions of subjective, insufficient information about objectively completely determined events. This is the very point of Popper's objective interpretation of physical indeterminism. The concept of emergence remains rather vague, as he admits (1977, p. 16), but this revision of the concept of causality allows for the interaction between structures on different levels. It combines chance and necessity, not additively but systematically, in one concept of emergent situational causality.

The concept of propensity implies (a) that indeterministic processes on a certain level may be influenced by configurations of parameters on higher (or lower) levels which are also indeterministic, (b) that the structure of laws on a higher level results from a stable dynamic regime which cannot be deduced from local dynamics on a lower level without at least some knowledge of the relevant parameters on its own and even higher levels, and (c) that indeterminism should not be interpreted as absolute randomness, but rather as a changing configuration of propensities which are bounded by the total hierarchical context of interaction. This concept comes close to other interpretations of modern physics, for instance by David Bohm, and to the emergent science of synergetics, as it crystallizes

in the works of Prigogine and Haken.[13] But to sociologists such theses
should sound quite familiar, since they remind us of Durkheim's
characterization of the relative autonomy of sociology.

IV. PROPENSITIES AS COLLECTIVE SOCIAL FORCES: DURKHEIM

A synthesis of the concept of propensity (Popper), emergent context
effects (Sperry) and the theory of dissipative structures (Prigogine)
should prove to be fruitful. Moreover, Prigogine himself has noted the
relevance of his thermodynamic and chemical structural models for the
problem of structural determination as discussed by Durkheim and
Althusser.[14] The theory of dissipative structures assumes that, under
conditions of disequilibrium, small forces may become decisive for the
development of stable structures on higher levels. These structures are
produced by global rather than local conditions. They depend on some
autocatalytic processes or positive and negative feedback mechanisms which
produce and reproduce gradients of unequal distribution of elements,
energies etc. Changing global conditions (i.e., rising levels of energy
supply) may induce a bifurcation or a "catastrophe" in system organization
and lead to qualitatively new global structures.

In developing his "social physics", Durkheim did not have these models
at his disposal, but his preference for the analogy of chemical synthesis
and his interest in statistics are well known. It is remarkable how he
explains (in his work on the division of labour) the importance of the
number of people and their internal lack of structure, in the sense of
their adaptability to arbitrary social positions and functions, as the
essential factors for the emergent complexity and instability of social
structures. Ideas and ideals gain their influence on the basis of these
unstable fluctuations which are receptive to comparatively small forces
inducing social order. This self-organization of society leads to the
development of higher forms of individuality, not the relatively stable
physical environment, as interpreters of Durkheim as an environmental
determinist and functionalist would claim. Moreover, as Durkheim argues
against Spencer (and could have argued against Popper), the
conceptualization of society as an **instrument** of individuals in adapting
to the physical world would explain only the limited equilibrium which so-

called primitive societies have attained. The selective pressure of Wl -
in Popper's terminology - does not explain social evolution.

Instead, Durkheim assumes a state of disequilibrium of social processes
continuously producing new propensities and structures. In these processes
the concrete, material and social milieu is a relatively stable base of
conditions, collective sentiments or emotions are the motoric, energetic
element, and ideas are the structure-inducing variables or "order
parameters" (Haken). The ideas are objectified and stabilized by means of
an external symbolism (language, writing, cultural objects) and function
as "catalysts" of social practices.   Thus, Bohannan (1960) compares
Durkheim's concept of collective consciousness to a force field
established by the interdependence of ideas.[15] In his socio-ecological
approach Durkheim succeeds, furthermore, in incorporating into the social
"situational logic" (Popper) the historical contingencies, demographic
developments and constraints of increasing social scale.

The concept of collective consciousness resembles rather closely
Popper's W3. This holds particularly for the objective, real existence of
traditions independent of specific individuals, for the importance of the
material world as a vehicle of W3 objects, i.e. the technological and
cultural products of men, and for the realistic interpretation of social
institutions.

In the present context, the parallels between Popper's theory of the
body-mind relationship and Durkheim's article on the same topic published
in 1889 (Durkheim, 1953) are of particular interest. Durkheim objects with
similar arguments to an "atomism" which somehow adds unchanging psychic
elements to psychic states by means of association. He holds panpsychism
to be as untenable as any other strategy ascribing properties of higher
levels to their elements as dispositions. Like Popper, he proposes as a
central argument the thesis that, in an evolutionary framework at least,
we have to assume the causal interaction of phenomena and that therefore
epiphenomenalism does not make sense. Life is not an epiphenomenon of
anorganic matter and, by the same token, of biologico-physiological
processes. An implication which Durkheim seizes upon and which Popper does
not discuss, at least not explicitly, is the applicability of this
argument to the relationship of psychic and social phenomena. Why should
individual psyches not give rise to the emergence of collective ideas

quite analogously to the emergence of individual ideas from neural structures?

The collective consciousness is a theoretical entity for Durkheim, but, like all theoretical entities, it is interpreted as real, although unobservable on the level of the individual. This becomes perfectly clear in his discussion of suicide (1951). Durkheim poses the problem of explaining a **tendency** or **collective disposition** for the occurrence of suicides. He explicitly seeks to explain not the causes of individual acts, but their situations, or the properties of their social context. That is why he chooses the suicide **rate** as an indicator which, in his judgment, is essentially independent of factors of the physical environment and psychological dispositions, and thus should serve as a valid measure of the collective social fact. The emergence of such social tendencies as social facts occurs on the basis of a concrete "inner milieu" - an objective infrastructure of individuals, their products and their environments - just as psychological phenomena arise in the context of the physiological structure of organisms.

Thus Durkheim is much more explicit about the socio-ecological structure of those situations which Popper employs in his "situational logic" to explain human actions and the growth of knowledge. If ideas, theories or, more generally, W3 objects are embodied as a rule in W1 objects and if "we operate with these objects almost as if they were physical objects" (Popper, 1972, p. 163), then it follows that the problem situation of a theory has an historical, concrete infrastructure. The propensity of an historical situation to favour the emergence of a theory (at least in the sense of the probability of being selected and accepted, if invented) has a material dimension. The problem situation of a theory depends on a stratification of knowledge which is not independent of the material embodiment of W3 objects. The implications of this line of thought for an interdependence between the theory of knowledge and the sociology of knowledge - an interdependence which Popper always fought tooth and nail - become obvious.

Now Durkheim has been justly criticized for his attempt to explain the concepts, myths and theories of social groups by their social structure in a straightforward fashion.[16] But this does not invalidate his theses on the importance of externalized symbols and collective emotionality in the emergence of the social sphere. His theory of symbols as "catalysts" in the coordination of social action does not draw a sharp line between the cognitive and the emotional aspects of ideas.

"A representation is not simply a mere image of reality, an inert shadow projected by things upon us, but it is a force which raises around itself a turbulence of organic and psychical phenomena". (1964, p. 97)

And this "turbulence" reaches beyond the individual and produces structures of collective sentiments and ideas in the interaction process. These structures exert a plastic control - in Popper's terminology - which, because of the externalization of symbols and the interaction effects in social processes, is relatively stable against deviations or fluctuations at the level of human behaviour. Thus, we could interpret Durkheim's concept of social tendencies as a propensity interpretation of social phenomena. Incidentally, Popper himself pointed out the relationship between preferences or dispositions and propensities (1974a, p. 143). If these social propensities are not to be interpreted subjectivistically or within the framework of an individualistic action theory, we have to place the symbols or W3 objects as the relevant "catalysts" in a socio-ecological context.

The structured interdependence of individual actions, their self-organization in more or less institutionalized, collective patterns, and the function of ideas can be modelled by synergetic structures and processes, at least in principle, since reasonably valid models will probably turn out to be intractable. For Durkheim such models were not available, and this is one of the reasons he tends to neglect, in his "mechanics of solidarity", the interdependence of different levels. Moreover, as Durkheim recognizes in his criticism of Spencer, the contemporary thermodynamics of equilibrium does not furnish a suitable model for creative, emergent structural change. Thus he uses the statistical mechanics of his day primarily in the tradition of Quetelet, i.e., as an argument for the relative causal independence of global structures from local micro dynamics.

A Durkheimian interpretation of Popper's W3 may now suggest itself, but some precautionary remarks are in order. First, Durkheim explicitly accepts only two "worlds". His solution to the cutting problem runs horizontally, as it were, through the "homo duplex" and distinguishes the psychophysical organism, as contemporary psychology described it, from the social self conceived as an integral element of collectivities (Durkheim, 1960).

The individual **is** the interface in Durkheim's conception. He is the product of individuating processes on the biological and the social level. While Popper places the interaction effects of W1 and W3 in a world of their own, for Durkheim this world of individual actions is nothing but the process of socio-cultural interaction in a physical environment. Although Durkheim describes an historical and social process of the evolution of individualism within society, he does not introduce the individualistic concept of a personal self prevalent in today's psychology and action theory.

One of the reasons for this difference between Durkheim and Popper is based on their theories of morality. Popper (1977, p. 145) sides with Kant in appealing to the moral responsibility of the individual, who uses his socio-cultural environment (although not other people in their capacity as moral agents) instrumentally in his moral acts. For Durkheim, the moral quality of a social milieu is expressed in individual actions, and individual responsibility depends on the capacity of the collectivity to organize responsible actions in "organic solidarity" and to implant the competence for responsible cooperation through socialization. Durkheim conceptualizes individual acts as tokens of social types; their creativity (**vs.** individualistic deviactions and fluctuations) belongs to the social context of interaction. Creativity is not the expression of an individual personality, as it is for Popper.

Durkheim has conceptualized the dualistic interface more adequately than Popper, who rather blurs it by his W2. Interestingly, it seems that a fresh reading of Durkheim's sociology of religion - along with criticism by Lidz and Lidz (1976) - led Parsons to acknowledge some inadequacies in his own distinction between organisms, actors and social systems. As a result, Parsons reorganized his framework and introduced the differentiation of non-symbolic from symbolic processes as the decisive interface of interpenetration in his cosmological hierarchy (Parsons, 1978, pp. 213-214 and p. 353).

Whether one follows Parsons in his Freudian framework for the analysis of the relation between these two levels or not, we have in Parsons' scheme a dualism which evidently distinguishes the interaction between these two levels from the interactions within biological and social levels respectively. We have already suggested the assumption of a distinction between inter- and intralevel emergence in Popper's framework. Following Parsons' terminology, we may distinguish between the **interpenetration** of relatively autonomous systems and the **interaction** between subsystems. Now, the paradigm of the self as a parasite provides a suggestive analogy for dualistic interpenetration, the concept of propensity redefines structural relationships to allow for "downward causation", and the theory of dissipative structures could furnish the models for a better understanding of interaction processes between levels.

V. THE SELF AS A PARASITE

A prerequisite for a biological approach to dualism is that the unity of the individual can be differentiated into parts. Thus, Popper argues against Strawson's thesis assuming the primacy of the concept of person over any body-mind differentiation (1977, pp. 115-116). The unity of the self is rather precarious and not a self-evident basis of experience, according to Popper. It develops ontogenetically and is not guaranteed by a pre-existing "pure ego" (1977, p. 111). To argue for the multiplicity of relatively independent mental subsystems and a relative independence of the self from the brain, Popper refers to research on "split-brain" patients and to speculations on brain transplantations.[17]   An earlier version of this pluralistic approach appears in Poppers's thesis on "genetic dualism" - or "genetic pluralism", as he calls it in an addendum, apparently after a recount (1972, p. 281). There he distinguishes between three subsystems, those of physiological structures, behavioural dispositions and preferences or goal structures, which exert relatively independent selection pressures on each other.

The "genetic dualism", however, does not justify a Cartesian dualism of self and brain, since gene structures only establish an "upward" causation. More adequate would be the conception of the entire genetic structure as a kind of "cuckoo's nest" (including robin **etc.**) with the self as the foisted egg. To support the analogy, we should point out that symbiotic or parasitic relationships may exist between organisms of different levels. In the "architecture of complexity" (Simon 1969), higher levels of organization can quite generally be said to exist parasitically on lower levels, since the independent survival of lower levels is a precondition for further evolution (Eckberg, 1981).[18]   Recall, additionally, that the distinction between species in symbiotic or parasitic relationships requires a reference to different ecological niches. The common "fate" of a society of genes in an ecological contest is, in a certain sense, a more fundamental basis for its unity than its physiological structure.

Now which criteria does Popper propose for his quasi-ontological cuts between W1 and W2 on the one hand, and W2 and W3 on the other? According to the scheme of "cosmic evolutionary stages" (1977, p. 16), the emergence of consciousness constitutes the new quality of W2 versus W1, and certainly Popper refers repeatedly to consciousness in arguing for the relative autonomy of W2. But his various remarks on consciousness do not combine into any coherent account. Now we have to admit that there is no satisfactory theory of consciousness available. Rather, we observe a progressing dissolution of the problem into a variety of psychological phenomena to which the labels "conscious" or "consciousness" are applied (Natsoulas, 1981). But this is precisely why consciousness becomes a rather questionable criterion for distinguishing worlds.

A closer look at Popper's account reveals the ambiguities. First, the aspect of subjective particularity and unity of experience is not constitutive for W2, since Popper apparently wants to distinguish a cosmological stage or level which is a psychological or mental realm of conscious beings, but which is not essentially private. Popper counts higher animals among these conscious beings, since they seem to employ his general strategy of trial-and-error, i.e., show intelligent and deliberate behavior (1977, pp. 120-129). To be conscious means, in this context, to direct attention actively and selectively to certain aspects of a situation, to choose among alternative courses of behaviour, and to anticipate behavioural outcomes. The problem is that Popper suggests that classifying intelligent behaviour is a rather simple task. As he states in a well-known phrase, there is trial and error and learning in behaviour from the amoeba to Einstein. To characterize consciousness as an "organ" for the solution of non-routine problems is not particularly revealing if

you cannot provide a more precise concept of non-routine problems to begin
with (1977, pp. 125-129). Since Popper tends to stress the gradual and
multidimensional development of intelligent behaviour (p. 123), he ends up
with questionable formulations about the identity of trial-and-error
behaviour and conscious or deliberate behaviour (p. 12). In a similar way,
he makes inflated use of the label "active" rather than "passive" with
regard to living phenomena. Thus, the distinction becomes useless in
characterizing consciousness, and we are left with no precise criterion
for ontological distinctions.

A different concept of consciousness, in the sense of being a possible
object within the limited span of present attention, pertains to the
distinction between conscious and unconscious process. In his sketch on
the development of the self, Popper does distinguish between attention and
consciousness, between (sub-)conscious and principally unconscious
processes, and between conscious and unconscious motives of decisions. But
he introduces the concepts not systematically, and without reference to a
distinction of W2 from W1. Popper obviously is not concerned with
selective attention, but rather with the unity of psychological processes
constituting the experience of a continuous self (1977, pp. 129-132). Even
language and rational criticism include mental processes - a "knowing how"
versus a "knowing that" - which cannot be elevated to conscious
introspection, as Popper points out (1977, p. 122). In fact, these
psychological processes, at least in humans, always seem to involve an
interaction between all three worlds, which makes the concept of an
interaction **within** W2 (Popper, 1977, p. 122) rather mysterious.

Obviously Popper's concept of mental processes in too broad to
delineate W2, since it includes fundamentally unconscious processes; the
actual world of the "here-now" of attention is too narrow, since it
excludes relevant aspects of the continuity of the self. Speaking of
potentially conscious dispositions raises the question of whether this
includes only those representations of our conduct and personality which
we can articulate in language (Searle, 1969, pp. 34-35). According to this
version, the W2 would be nothing but the process by which W1 mechanisms of
the brain work on socioculturally acquired W3 representations. Basically,
this is the model of cognitivism and philosophical functionalism, which
distinguishes only between W1 ("hardware") and W2/W3 ("software"). Again

we have a version of Durkheim's "homo duplex", and the dualistic distinction could be made more precise by means of Pylyshyn's (1980) methodological criterion of "cognitive impenetrability".

Another **biological** approach to consciousness is not considered by Popper. The sociologists Plotkin and Odling-Smee (1981) describe an important evolutionary step in their fourlevel model of evolution. With a motivation akin to Popper's interest in a theory of knowledge, they consider the following mechanisms of information storage developed in the course of evolution: genes, epigenetical adaptations, individual learning and memory, and social learning by traditions ("cultural pool"). Individual learning as a strategy of adapting to environmental changes too rapid for genetic adaptation appears very early in the evolution process. The crucial step leading to qualitatively new phenomena occurs with the emergence of traditions, independent of the survival of specific individuals. Such traditions do not require language and are thus not limited to humans. Moreover, the damping effects of traditions on the selection-pressures inducing adaptive changes in genetic structures can be observed in higher animals, too (Maynard Smith, 1982). An obvious exception to these damping effects are the adaptive changes in the genetic basis of the brain, which evolves into a general system for the processing of such traditions. A comparison of the position of Plotkin and Odling-Smee with that of Lumsden and Wilson (1982) reveals that, in the former, we have a classical case of "downward" causation. In the latter, Lumsden and Wilson introduce, as Loftus (1982) criticizes, another causal and selective influence "from below" through their concept of "culturgenes". which can hardly explain individual and cultural variation in cultural objects.

Popper's position appears to be rather ambiguous: "Cultural evolution, we may say, continues genetic evolution by other means: by means of World 3 objects" (1977, p. 48). But first, cultural evolution is a new emergent process and certainly not just a new means of genetic evolution. Popper himself stresses that human culture is determined by our genetic endowment only in a very general way. Second, cultural evolution, specifically the growth of knowledge, does not progress under some instrumental or pragmatic principle of usefulness and, is therefore, **a fortiori** not in close functional relationship with the evolution of gene structures. Thus,

Popper should emphasize the dualistic and interactionistic nature of coevolution much more than he actually does.

The question arises, why does Popper describe the creative evolution of worlds within such a narrow biologistic framework. He even attributes to some individual instinct (!) of survival a rather central role in the unity of the self (1977, p. 127). One reason seems to stem from unresolved problems in the delineation of worlds. Obviously, a coevolution of genes and memes corresponds rather well to a parasite theory of self, as we attribute it to Popper. But then we would have only two worlds, W1 and W3. Moreover, W3 would extend down to the level of animal behaviour, since we find traditions of learned behaviour on this level, too. For Popper, emphasizing the biological basis of individual capacities apparently serves to push up the boundary between W3 and the other two worlds, but it does little to clarify the nature of W2.

Now, the particularity of W3 objects indicates that there is a relevant difference between social learning, in the sense of animal learning traditions, and social or cultural learning in humans. As Anderson (1961; 1962) notes, the very definition of cultural objects depends on a concept of "learning-from-another-human-being". But since Popper disregards the social science literature almost entirely, he is ill equipped to handle this problem. His favorite paradigm, trial-and-error learning or problem-solving, just does not help. There is an essential difference between more or less deliberate choices among behavioural alternatives, and taking the deliberate choices of other actors into account. Neither individual agency as opposed to passivity nor the unity of experience established by our conscious or unconscious monitoring of a continuous behaviour stream are sufficient for the emergence of mental processes in the sense of Popper's subjective W2. Such a world does emerge, when individuals are capable of "taking the role of the other" (Mead, 1934) and attribute different subjective perspectives on the situation to other actors. Subjectivity in this sense arises together with and in contrast to intersubjectivity and objectivity, i.e. the ability to distinguish between the subjective roles and an "observer's" perspective on the interaction. Certainly Popper emphasizes the importance of language, of learning pronouns and of an "animistic" disposition of children to attribute personhood to objects. But Popper stays mute on the implications of learning pronouns and of interpreting other persons as mirrors of one's own personality (1977, p. 49 and p. 110). If you like, the emergence of W2 occurred because, at some point, our biological ancestors preferred not to remain tough behaviourists, but to act under the shared supposition of a subjective perspective of each partner to some interaction.

Popper's remarks on the simultaneous emergence of human subjectivity and language point to the importance of the development of symbolic representations. However, he misses the essentially intersubjective nature of symbols, which are not just ambiguous indicators of future states of the world, but imply alternative interpretations of the world from different perspectives. As Popper notes in the dialogues with Eccles, the remarkable fact in cultural evolution is that the entire environment acquires the character of meaningful symbols. There emerges a new type of environment or a new ecological niche. Speech and written language are only specific developments of the primary capacity for semiosis. Moreover, not only does a medium of communication develop but, in the same process, the self emerges as a perspectival "location" within the network of symbolic interaction: a "parasite" taking hold of the central nervous system and introducing a new world of perspectives and aims into the structure of behaviour.

Again, this evolutionary or emergent step establishes only two worlds, i.e., symbolic and non-symbolic processes. To speak of two worlds in this case appears plausible, since it supports the relevant distinction between inter-subjective and objective states of affairs, between meaning and reference, and between true and false theories. The subjective worlds of selves are generated in a symbolic interaction process, as Mead has shown. Moreover, from Durkheim we may learn that the individuality of selves is largely dependent on a social process which differentiates roles in such a way that taking the role of others will actually result in a wealth of different perspectives which, at the same time locate the self in a very particular position in the social context.

But, by its very contrast, a Durkheimian approach reveals another feature of Popper's W2. In defending the autonomy of W2, Popper is actually defending the autonomy of a liberal individual self and equating these two concepts of autonomy. Popper argues for the difference between a biological W1 and a social W2 with reference to the **individual** survival

of a personal self which knows about its own continuity as a self and
about a personal death. Human beings have a self-consciousness, not only
consciousness. But, as Popper himself states, this self-consciousness
consists of a relationship between consciousness and a kind of **theory**
about one's own self (1977, p. 145). Since consciousness is supposed to
hold for higher animals too, and since theories are paradigmatic cases of
W3 objects, we are again left with an interaction of W1 and W3 with the
facts of individual existence as the object of a symbolic process.

VI. EPISTEMOLOGY AND THE KNOWING SUBJECT

The discussion shows that there is another sense in which the self appears
as a parasite in Popper's theory. Preconceptions about the self induce him
to tolerate a strange and inconsistent element within his dualism, that of
a parasite sapping the strength of his cosmological theory. This parasite
is World 2!

As I see it, for two equally unsatisfactory reasons Popper does not
restrict his declared dualistic position to an interaction of W1 and W3:
First is his distinction between subjective and objective knowledge and
second is the role he assigns to the self as an ethically and morally
responsible actor.

Now, the epistemological distinction between subjective and objective
knowledge is irrelevant in a cosmological context. Popper argues for the
existence of objective knowledge and against some rather unspecified
"subjectivistic" or "sociology-of-knowledge" approach which treats
scientific theories as **nothing but** subjective opinions depending on the
particular psychological makeup of individuals. Surely Popper is justified
in pursuing his theory of objective knowledge. In this respect there is no
alternative to a modern and modified version of Platonism or even
Hegelianism, as Popper characterizes his approach (1972, pp. 122-126).
Anderson (1961; 1962) argues convincingly that any satisfying theory of
sociocultural phenomena will be a Platonism of sorts. Moreover, if we have
to revise Platonism, we certainly have to incorporate, as Popper does, a
new theory of time and the emergent evolution of cultural objects.
Whitehead would appear to be a promising reference in this respect, but he
is barely mentioned by Popper. In any case, the creativity of individual

selves is not enough, to paraphrase Popper's dictum on indeterminism, for a theory of the **production** and **selection** of cultural objects. Looking at it from a sociological perspective, we should recognize that "subjectivism" is not only untenable in epistemology, but is also no good in the social sciences.

Thus the problem lies not with Popper's arguments for W3, but with his introduction of W2. Certainly, it was an important step in Popper's logic of discovery to declare all knowledge hypothetical, even subjective evidence. It is the very point of the "epistemology without a knowing subject" (1972, pp. 106-152) that it is **irrelevant** in which forms our knowledge is "embodied". But this applies equally to embodiment in empirical subjects; there is no need for ontological distinctions. Surely it is only "human" to attribute a special relevance to our own personal knowledge. After all, it is not only a means of communication, but also part of our personalities and world views. But Popper's ambiguous remarks on the causal interactions of W2 and W3 "grasping" W2 objects, however, suggest some kind of ontological metamorphosis or copying of symbolic contents which enables them to appear in the subjective world (1977, p. 44). As to the individuality of the self, it is granted by the distinct, perspectival position in the context of symbolic interaction. And particularly, as Popper himself states, there is no "pure ego" beyond the structure of this context (1977, p. 111).

On the other hand, criticism of Popper's W3 that opposes his identification of historical and sociocultural objects with the timeless and abstract realm of theories or even mathematical structures is also misguided.[19] We may grant that the **philosophical** problems of Plato can be solved. But we do not have an **empirical** theory of symbolic processes which would allow us to draw such sharp distinctions between sociocultural and abstract realms. What we observe, instead, is a rising complexity of social theories with growing tendencies towards "cognitivism", especially in psychology and the theory of natural languages. The best psychological theories we have at present may well be incorporated in mathematics and formal logic.

Another misleading interpretation of W3 rather common among sociologists is prompted by Popper's distinction between subjective and objective knowledge. Thus, Habermas (1981, vol. 1, p. 119) and Giesen

(1980, p. 17), arguing from rather different sociological perspectives, both take W3 refer to (more or less) scientifically justified or rational sociocultural stocks of knowledge as opposed to individually held or socially shared norms and beliefs. Now, I would agree that there is a lot of merit in distinguishing stocks of knowledge with reference to explicit or implicit strategies of justification (Pieper, 1979). These provide, for instance, criteria for a sociological distinction between subjective and objective knowledge. But Popper's W3 is not objective in **this** sense; it rather encompasses all sociocultural objects, **i.e.**, ideologies, prejudices, **false** theories, objects of art and music, instrumental knowledge, technological principles, rituals, institutions, religions etc. Thus, W3 constitutes the domain of cultural anthropology, not the realm of some rationally or scientifically exceptional knowledge. This is one of the reasons W2 is a "parasite" in Popper's scheme! Habermas and Giesen do not pose Compton's problem at this point; they both accept a version of the dualism of non-symbolic and symbolic processes including the self as a **social** actor among sociocultural phenomena. They only try to make use of Popper's distinction between W2 and W3 **within** the sociocultural domain.

Let me close with a final word regarding Popper's concern about the responsibility of the ethical-moral subject. We have pointed out that Popper disrupts the relationship of individual and society by the way he divides the universe into worlds. Only within an irresponsible society does the individual have to bear responsibility to the extent visualized by the rationalist and liberalist Popper. The sociocultural background, even of individual value systems and moral conduct, should persuade Popper to adopt a concept of ethical-moral action which does not implicitly **force** the individual into "existentialistic" creativity in the realm of ethics in order to cope with an amoral social environment. The individual actor is not the only source of morality in a social world.

NOTES

1. Popper (1977, p. 105) alludes explicitly to the parallels between his dualism and the idea of a ghost in a machine criticized by Gilbert Ryle in "The Concept of Mind" (1949).

2. Popper's trialistic theory has received a rather mixed reception, ranging from praise to harsh criticism. One reason for this may be, as

Dennett (1979) notes, Popper's neglect of almost the entire contemporary psychological and philosophical discussion on the body-mind problem. From a sociological point of view, it is equally irritating that Popper assumes a central role for cultural objects in his theory without mentioning any literature in sociology or social philosophy. For discussions of Popper's theory, see Beloff (1978), Bunge (1977), Carr (1977), Currie (1978), Dennet (1979), Grove (1980), Jackson (1980), MacKay (1978), Mackie (1978), Sperry (1980) and Werth (1980). For the contemporary discussion of the body-mind problem, see for instance Putnam (1982), Block (ed.) (1980; 1981), Dennett (1978), Globus, Maxwell and Savodnik (eds.) (1976) and Rorty (ed.) (1976).

3. See, for instance the article by Beloff (1978).

4. Popper describes his changing judgment on Darwin's theory of evolution in his autobiography (1974a; see also 1978a).

5. Popper withdraws this thesis, however, in a later article (1978a). For criticism on this point see Jackson (1980).

6. The theory of self and consciousness poses a variety of problems which Popper leaves aside. Natsoulas (1981) furnishes a good survey of the aspects of these problems as currently discussed.

7. As Settle (1981) points out in a criticism of Bunge, his materialism merely places the problem one step lower in the hierarchy of organized matter. Bunge now has to explain instead the relative autonomy of neural processes from physical and chemical processes.

8. Popper refers to the computer analogy in an earlier article (1972, p. 248 ff). That the discussion of the mind-brain relationship from an "artificial intelligence" perspective can be at once profound and entertaining is demonstrated by Hofstadter and Dennett (Hofstadter 1979, Dennett 1978, Hofstadter and Dennett (eds.) 1981). See Hofstadter and Dennett (eds.) 1981, pp. 465-482, for a guide to the literature on this and other approaches to the mind-body problem.

9. This is not quite correct, since interacting stochastic systems can simulate deterministic behaviour.

10. For a discussion of autopoiesis see Zeleny (ed.) (1981).

11. "It is the idea, in brief, that conscious phenomena as emergent functional properties of brain processing exert an active control as causal determinants in shaping the flow pattern of cerebral excitation. Once generated from neural events, the higher order mental pattern and programs have their own subjective qualities and progress, operate and interact by their own causal laws and principles which are different from and cannot be reduced to those of neurophysiology, as explained further below. Compared to the physiological processes, the conscious events are more molar, being determined by configurational or organizational interrelations in neuronal functions. The mental entities transcends the physiological just as the physiological transcends the molecular, the molecular, the atomic and subatomic, etc. The mental forces do not

violate, disturb or intervene in neuronal activity but they do supervene."
(Sperry, 1980, p. 201) Sperry characterizes his position as "emergent
determinism" to emphasize the lawfulness of emergent processes. Since he
accepts indeterminism in physics, his position does not seem to be
different from Popper's on this point. For an interesting discussion on
emergent levels of organization in the brain, see Puccetti and Dykes
(1978) and the commentaries by Bridgeman, Dennett, Mandler, Maxwell and
Szentagotai; see also Dewan (1976).

12. See Eccles' contribution in "The Self and Its Brain", especially the
dialogues with Popper in the third part of the book. See also Eccles
(1974) and (1976).

13. See Bohm (1981), Prigogine (1976; 1981) and Haken (1978).

14. Prigogine (1976, p. 122); cites Althusser and Balibar (1973, p. 61)
and refers, among others, to Durkheim and Quetelet. The connections
between Durkheim and thermodynamics suggest the development of a "new
social physics" in the light of these modern advances in synergetics, as I
will argue elsewhere (forthcoming).

15. Such physical and chemical analogies in Durkheim's own work are not
just unfortunate illustrations as some interpretations suggest (König,
1973, p. 498). The common depictions of Durkheim as a functionalist and
action theorist completely miss the point that he develops his paradigm of
social physics in opposition to Spencer and Tarde and within the framework
of a natural philosophy. Moreover, his realism leads Durkheim (1966) to a
philosophy of science which is very close to Popper's logic of discovery -
and far from positivism which is often attributed to him.

16. See, for instance, the introduction to Durkheim and Mauss (1963) by
Needham.

17. "Split brains" result from a physiologically or pharmacologically
induced division of the two hemispheres of the brain. An interesting
experiment on the multiplicity of mental systems in the brain is described
by Le Doux et al. (1979). Brain transplantations provide an attractive but
questionable **gedanken** experiment on the brain-mind relationship. First,
the question as to which **common** characteristics of the the donor and
recipient have to be presupposed for a successful experiment is generally
evaded. For instance, how does the brain of a Spanish flamenco dancer
recovering consciousness in the body of a German cleaning woman interpret
his/her menstrual cramps? Second, there is no discussion of the social
interaction context in which relevant aspects of the **new** identity of the
patient are developed and reinforced. What if a male-chauvinist doctor and
a feminist nurse each reinforce the continuities with a different gender
identity (see also Rorty (1976) and Settle (1981).

18. Some authors even assume the existence of parasitic genes f. e.
Perzigan (1981), and Cloak (1981).

19. For a critical discussion of Popper's W3, see Beloff (1978), Carr
(1977), Currie (1978), Grove (1980), Mackie (1978) and Quinton (1973).

BIBLIOGRAPHY

Abercrombie, N.: 1972,' 'Evolutionary Theories Again', **Sociological Analysis**, 2, pp. 47-51.

Adams, R. N.: 1975, **Energy and Structure. A Theory of Social Power**, University of Texas Press, Austin-London.

Agassi, J.: 1977, **Towards a Rational Philosophical Antropology**, Martinus Nijhoff, The Hague.

Albert, H.: 1967, 'Modell-Platonismus. Der neo-klassische Stil des ökonomischen Denkens', in H. Albert, **Marktsoziologie und Entscheidungslogik**, Luchterhand, Neuwied, pp. 331-367.

Aldrich, H. E.: 1979, **Organizations and Environments**, Prentice Hall, Engelwood Cliffs.

Alexander, R. D.: 1979, **Darwinism and Human Affairs**, University of Washington Press, Seattle.

Alland, A.: 1980, **To Be Human**, J. Wiley, New York.

Almond, G. A.: 1973, 'Approaches to Developmental Causation', in G. A. Almond et al. (eds.): **Crisis, Choice und Change: Historic Studies of Political Development** Little, Brown, Boston, pp. 1-42.

Almond, G. A., and J. S. Coleman, eds.: 1960, **The Politics of the Developing Areas**, Princeton University Press, Princeton.

Althusser, L., and E. Balibar: 1973, **Lire le Capital**, Maspero, Paris.

Anderson, A. R.: 1961, 'What do Symbols Symbolize Platonism', **Philosophy of Science**, pp. 137-151.

Anderson, A. R. and O. K. Moore: 1962, 'Toward a Formal Analysis of Cultural Objects', **Synthese**, 14, pp. 144-170.

Andreski, S.: 1968 (1954), **Military Organization and Society**, University of California Press, Berkeley and Los Angelos.

Ashby, W. R.: 1956, **An Introduction to Cybernetics**, John Wiley, New York.

Axelrod, R.: 1981, 'The Emergence of Cooperation Among Egoists', **American Political Science Review**, 75, pp. 306-318.

Axelrod, R., and W. D. Hamilton: 1981, 'The Evolution of Cooperation', **Science**, 211, pp. 1390-1396.

Barash, D.: 1978, 'Evolution as a New Paradigm for Behavior', in M. S. Gregory and A. Silvers (eds): **Sociobiology and Human Nature. An Interdisciplinary Critique and Defense**, Jossey-Bass, San Francisco, pp. 13-33.

Barash, D.: 1980, **Soziobiologie und Verhalten**, P. Parey, Berlin.

Barash, D.: 1981, **Das Flüstern in uns**, S. Fischer, Frankfurt.

Barash, D.: 1982, **Sociobiology and Behavior**, 2nd ed., Elsevier, New York.

Bartley, W. W. III: 1976, 'Critical Study: The Philosophy of Karl Popper: Part I: Biology & Evolutionary Epistemology', **Philosophia**, 6, pp. 463-494.

Bartley, W. W. III: 1983, 'The Challenge of Evolutionary Epistemology', **Proceedings of the Eleventh International Conference on the Unity of the Sciences**, vol. 11, International Cultural Foundation Press, New York, pp. 835-880.

Bateson, G.: 1982, **Geist und Natur. Eine notwendige Einheit**, Suhrkamp, Frankfurt.

Baum, R. C.: 1976, 'On Societal Media Dynamics', in J. J. Loubser, R. C. Baum, A. Effrat, and V. M. Lidz (eds.): **Explorations in General Theory in Social Science**, The Free Press, New York, pp. 579-608.

Beloff, J.: 1978, 'Is Mind Autonomous?', **British Journal for the Philosophy of Science**, 29, pp. 265-283.

Berger, P. L. and T. Luckmann: 1966, **The Social Construction of Reality: A Treatise in the Sociology of Knowledge**, Doubleday & Co., New York.

Bertalanffy, L. von: 1949, **Das biologische Weltbild**, Francke, Bern.

Bertalanffy, L. von: 1973, **General System Theory: Foundations, Development, Applications**, Penguin University Books, Harmondsworth.

Blau, P. M.: 1977, **Inequality and Heterogeneity. A Primitive Theory of Social Structure**, The Free Press, New York-London.

Bloch, M.: 1982, 'The Disconnection of Power and Rank as a Process. An Outline of the Development of Kingdoms in Central Madagascar', in J. Friedman and M. J. Rowlands (eds.), **The Evolution of Social Systems** (2nd. ed.), Duckworth, Gloucester Crescent.

Block, N. (ed.): 1980, **Reading in the Philosophy of Psychology**, 2 vols., Harvard University Press, Cambridge, Mass.

Blute, M.: 1979, 'Sociocultural Evolutionism: An Untried Theory', **Behavioral Science**, 24, pp. 46-59.

Boehm, C.: 1978, 'Rational Preselection from Hamadryas to **Homo Sapiens**: The place of decisions in adaptive process', **American Anthropologist**, 80, pp. 265-296.

Bohannan, P.: 1960, 'Conscience Collective and Culture', in K. H. Wolff (ed.), **Emile Durkheim: 1858-1917**, Ohio State University Press, Columbus, Ohio, pp. 77-96.

Bohannan, P.: 1963, **Social Anthropology**, Holt, Rinehart & Winston, New York.

Bohm, D.: 1981, **Wholeness and the Implicate Order**, Routledge & Kegan Paul, London.

Boudon, R.: 1979, **Widersprüche sozialen Handelns**, Luchterhand, Neuwied-Darmstadt.

Boudon, R.: 1980, **Die Logik gesellschaftlichen Handelns**, Luchterhand, Neuwied-Darmstadt.

Boulding, K.: 1978, **Ecodynamics. A New Theory of Societal Evolution**, Sage, Beverly Hills - London.

Bourdieu, P.: 1979, **Entwurf zu einer Theorie der Praxis**, Suhrkamp, Frankfurt.

Bourdieu, P.: 1982, **Die feinen Unterschiede. Kritik der gesellschaftlichen Urteilskraft**, Suhrkamp, Frankfurt.

Bridgeman, B.: 1978, 'Open Peer Commentary: The Similarity of the Sensory Cortices: Problem or Solution?', **The Behavioral and Brain Sciences**, 3, pp. 349-350.

Brown, R., and R. J. Herrnstein: 1975, **Psychology**, Little, Brown, Boston.

Buckley, W.: 1967, **Sociology and Modern Systems Theory**, Prentice Hall, Englewood Cliffs.

Buckley, W., (ed.): 1968, **Modern Systems Research for the Behavioral Scientist**, Aldine, Chicago.

Bunge, M.: 1977, 'Emergence and the Mind', **Neuroscience**, 2, pp. 501-509.

Bunge, M.: 1979a, 'The Mind-Body Problem, Information Theory and Christian Dogma', **Neuroscience**, 2, pp. 501-509.

Bunge, M.: 1979b, 'A Systems Concept of Society: Beyond Individualism and Holism', **Theory and Decision**, 10, pp. 13-30.

Burrows, J. W.: 1966, **Evolution and Society. A Study in Victorian Social Theory**, Cambridge University Press, Cambridge.

Campbell, D. T.: 1965, 'Variation and Selective Retention in Socio-Cultural Evolution', in H. R. Barringer, **et al.** (eds.): **Social Change in Developing Areas**, Shenkman, Cambridge. Mass., pp. 19-49.

Campbell, D. T.: 1974, 'Evolutionary Epistemology', in P. A. Schilpp
    (ed.), **The Philosophy of Karl Popper**, vol. 1, Open Court, La Salle,
    Ill., pp. 413-463.

Campbell, D. T.: 1976, 'On the Conflicts Between Biological and Social;
    Evolution and Between Psychology and Moral Tradition', **Zygon. Journal
    of Religion and Science**, 11, 167-208.

Candland, D. K., **et al.**: 1977, **Emotion**, Brooks-Cole, Montary.

Carneiro, R. L.: 1967, 'On the Relationship Between Size of Population and
    Complexity   of   Social   Organization',   **Southwestern   Journal   of
    Anthropology**, 23, pp. 234-243.

Carneiro, R. L.: 1970, 'Scale Analysis, Evolutionary Sequences, and the
    Rating of Cultures', in R. Naroll and R. Cohen (eds.): **A Handbook of
    Method in Cultural Anthropology**, Natural History Press, Garden City,
    pp. 834-871.

Carneiro, R. L.: 1973a, 'Structure, Function, and Equilibrium in the
    Evolutionism of Herbert Spencer', **Anthropological Research**, 29, pp.
    77-95.

Carneiro, R. L.: 1973b, 'The Four Faces of Evolution', in J. J. Honigmann
    (ed.): **Handbook of Social and Cultural Anthropology**, Rand Nally,
    Chicago.

Carneiro, R. L.: 1978, 'Political Expansion as an Expression of the
    Principle of Competitive Exclusion', in R. Cohen and E. R. Service
    (eds.): **Origins of the State: The Anthropology of Political Evolution**,
    ISHI, Philadelphia.

Carr, B.: 1977, 'Popper's Third World', **The Philosophical Quarterly**, 27,
    pp. 214-226.

Chase, R.: 1980, 'Structural-functional Dynamics in the Analysis of
    Socioeconomic Systems: Adaptation of Structural Change Processes to
    Biological Systems of Human Interaction', **American Journal of
    Economics and Sociology**, 39, pp. 49-64.

Chomsky, N.: 1980a, **Rules and Representations**, Columbia University Press,
    New York.

Chomsky, N.: 1980b, 'Rules and Representations (with open peer
    commentary)', **The Behavioral and Brain Sciences**, 3, pp. 1-61.

Claessen, H. J. M.: 1978. 'The Early State: A Structural Approach', in H.
    J. M. Claessen and P. Skalnik (eds.), **The Early State**, Mouton, The
    Hague-Paris-New York, pp. 533-596.

Claessen, H. J. M., and P. Skalnik: 1978a, 'The Early State: Theories and Hypotheses', in H. J. M. Claessen and P. Skalnik (eds.), **The Early State**, Mouton, The Hague-Paris-New York, pp. 3-29.

Claessen, H. J. M., and P. Skalnik: 1978b, 'The Early State: Models and Reality', in H. J. M. Claessen and P. Skalnik (eds), **The Early State**, Mouton, The Hague-Paris-New York, pp. 637-650.

Claessens, D.: 1980, **Das Konkrete und das Abstrakte. Soziologische Studien zur Anthropologie**, Suhrkamp, Frankfurt.

Clarke, B.: 1975, 'The Cause of Biological Diversity', **Scientific American**, 233, pp. 50-60.

Cloak, F. T. Jr.: 1981, 'Open Peer Commentary: On Natural Selection and Culture', **The Behavioral and Brain Sciences**, 4, pp. 238-240.

Cohen, R.: 1978, 'State Origins: A Reappraisal', in H. J. M. Claessen and P. Skalnik (eds.), **The Early State**, Mouton, The Hague-Paris-New York, pp. 31-75.

Cohen, R., and E. R. Service (eds.): 1978, **Origins of the State. The Anthropology of Political Evolution**, ISHI, Inc., Philadelphia.

Corning, P. A.: 1971, 'The Biological Bases of Behavior and Some Implications for Political Science', **World Politics**, 23, pp. 321-370.

Corning, P. A.: 1974, 'Politics and the Evolutionary Process', in T. Dobzhansky et al. (eds.): **Evolutionary Biology**, Plenum, New York, pp. 253-294.

Corning, P. A.: 1975, 'Toward a Survival-oriented Policy Science', in A. Somit (ed.): **Biology and Politics**, Mouton, Paris-The Hague, pp. 127-154.

Corning, P. A.: 1981, 'Durkheim and Spencer', **British Journal of Sociology**, 23, pp. 359-382.

Corning, P. A.: 1983, **The Synergism Hypothesis. A Progressive Theory of Evolution**, Macmillan, New York-London et. al..

Corning, P. A.: 1987, 'Evolution and Political Control. A Synopsis of General Theory of Politics', **this volume**.

Coulam, R. F.: 1977, **Illusion of Choice**, Princeton University Press, Princeton.

Count, E. W.: 1958, 'The Biological Basis of Human Sociality', **American Anthropologist**, 60, pp. 1049-1085.

Currie, G.: 1978, 'Popper's Evolutionary Epistemology: A Critique', **Synthese**, 37, pp. 413-431.

Darwin, C.: 1859, **On the Origin of Species**, John Murray, London.

Darwin, C.: 1968 (1859), **On the Origin of Species by Means of Natural Selection, or the Preservation of Favoured Races in the Struggle for Life**, Penguin, Baltimore.

Darwin, C.: 1871, **Descent of Man**, John Murray, London.

Darwin, C.: 1982, **Darwin. Ein Leben**, ed. by S. Schmitz, Deutscher Taschenbuchverlag, München.

Dawkins, R.: 1976, **The Selfish Gene**, Oxford University Press, Oxford.

Dawkins, R.: 1981, 'Selfish Genes and Selfish Memes', in D. R. Hofstadter and D. C. Dennett (eds.), **The Mind's I**, Bantam Books, New York, pp. 124-144.

Dennett, D. C.: 1978, **Brainstorms**, Bradford Books, Montgomery, Vt.

Dennett, D. C.: 1979, 'Book Review: The Self and its Brain', **Journal of Philosophy**, pp. 91-97.

Deutsch, K. W.: 1963, **The Nerves of Government: Models of Political Communication and Control**, The Free Press, New York.

Deutsch, K. W.: 1979, **Tides Among Nations**, The Free Press, New York.

Dewan, E. M.: 1976, 'Consciousness as an Emergent Causal Agent in the Context of Control System Theory', in G. G. Globus, G. Maxwelland J. Savodnik (eds.): **Consciousness and the Brain**, Plenum Press, New York, pp. 181-198.

Dougherty, J. E., and R. L. Pfaltzgraff, Jr.: 1971, **Contending Theories of International Relations**, Lippincott, Philadelphia.

Douglas, M.: 1978, **Purity and Danger. An Analysis of Concepts of Pollution and Taboo**, 2nd. ed., Routledge & Kegan, London.

Dumont, L.: 1967, **Homo Hierarchicus**, Presses Universitaires de France, Paris.

Dumont, L.: 1970, 'Religion, Politics, and Society in the Individualistic Universe', **Proceedings of the Royal Anthropological Institutions of Great Britain and Ireland**, 1970, 31-41.

Durham, W. H.: 1976, 'The Adaptive Significance of Cultural Behavior', **Human Ecology**, 4, pp. 89-121.

Durkheim, E.: 1951, **Suicide: A Study in Sociology**, Free Press, Glencoe, Ill.

Durkheim, E.: 1953, **Sociology and Philosophy**, Free Press, Glencoe, Ill.

Durkheim, E.: 1960, 'The Dualism of Human Nature and Its Social Conditions', in K. H. Wolff (ed.): **Emile Durkheim: 1858-1917**, Ohio State University Press, Columbus, Ohio, pp. 325-340.

Durkheim, E.: 1961, **The Elementary Forms of the Religious Life: A Study in Religious Sociology**, Collier, New York.

Durkheim, E.: 1964, **The Division of Labor in Society**, The Free Press, Glencoe.

Durkheim, E.: 1966, **The Rules of Sociological Method**, Free Press, New York.

Durkheim, E.: 1980, **Regeln der soziologischen Methode**, 6th printing, Luchterhand, Darmstadt.

Durkheim, E. and M. Mauss: 1963, **Primitive Classification**, Cohen & West, Chicago.

Easton, D.: 1965a, **A Framework for Political Analysis**, Prentice Hall, Englewood Cliffs.

Easton, D.: 1965b, **A Systems Analysis of Political Life**, Wiley, New York.

Eccles, J. C.: 1974, 'The World of Objective Knowledge', in P. A. Schilpp (ed.), **The Philosophy of Karl Popper**, Vol. I, Open Court, LaSalle, Ill. pp. 349-370.

Eccles, J. C.: 1976, 'Brain and Free Will', in G. G. Globus, G. Maxwell and J. Savodnik (eds.), **Consciousness and the Brain**, Plenum Press, New York, pp. 101-121.

Eckberg, D. L.: 1981, 'Open Peer Commentary: Some Problems with an "Options" View of Evolution', **The Behavioral and Brain Sciences**, 4, pp. 241-242.

Economos, J.: 1978, 'Open Peer Commentary: What Is It Like, Mr. Pucetti?', **The Behavioral and Brain Sciences**, 3, pp. 352-353.

Eder, K.: 1976, **Die Entstehung staatlich organisierter Gesellschaften. Ein Beitrag zu einer Theorie sozialer Evolution**, Suhrkamp, Frankfurt.

Eder, K.: 1984, 'On the Cultural Origins and the Historical Formation of the Traditional State: Some Theoretical Considerations', in W. Dostal (ed.), **On Social Evolution. Vienna Contributions to Ethnology and Anthropology**, Vol. 1, Ferdinand Berger, Horn-Wien, pp. 110-140.

Eder, K.: 1987, 'Learning and the Evolution of Social Systems. An Epigenetic Perspective', **this volume**.

Eigen, M., and P. Schuster: 1979, **The Hypercycle: A Principle of Natural Self-Organization**, Springer, Berling-Heidelberg-New York.

Eigen, M. and R. Winkler: 1981, **Das Spiel. Naturgesetze steuern den Zufall**, 4th printing Piper, München.

Eisenstadt, S. N.: 1963, **The Political Systems of Empires**, The Free Press, New York.

Eisenstadt, S. N., and M. Curelaru: 1976, **The Forms of Sociology. Paradigms and Crisis**, Wiley, New York-London-Sydney-Toronto.

Ekman, P.: 1979, 'About Brows: Emotional and Conversational Signals', in M. v. Cranach et al. (eds.): **Human Ethology. Claims and Limits of a New Discipline**, Cambridge University Press, Cambridge, pp. 169-203.

Elster, J.: 1978, **Logic and Society**, Wiley, New York.

Erikson, E. H.: 1966, 'Ontogeny of Ritualization in Man', **Philosophical Transactions of the Royal Society of London**, 251, pp. 337-349.

Fairservis, W. A.: 1975, **The Threshold of Civilization. An Experiment in Prehistory**, Scribner's & Sons, New York.

Fararo, T.: 1978, 'An Introduction to Catastrophes', **Behavioral Science**, 23, pp. 291-317.

Flannery, K.: 1972, 'The Cultural Evolution of Civilizations', **Annual Review of Ecology and Systematics**, 3, pp. 399-426.

Flew, A.G.N.: 1963, 'The Structure of Malthus' Population Theory', in B. Baumrin (ed.): **Philosophy of Science: The Delaware Seminar**, vol. 1, Wiley, New York, pp. 283 - 307.

Flew, A.G.N.: 1976, **Evolutionary Ethics**, Macmillan, London.

Francis, E. K.: 1981, 'Darwins Evolutionstheorie und der Sozial-darwinismus', **Kölner Zeitschrift für Soziologie und Sozialpsychologie**, 33, pp. 209-228.

Fried, M. H.: 1967, **The Evolution of Political Society. An Essay in Political Anthropology**, Random House, New York.

Fried, M. H.: 1975, **The Notion of Tribe**, Cummings, Menlo Park, Cal.

Friedmann, J., and M. J. Rowlands (eds.): 1982, 'Notes toward and Epigenetic Model of the Evolution of "Civilisation"', in J. Friedman and M. J. Rowlands (eds.), **The Evolution of Social Systems** (2nd. ed.), Duckworth, Gloucester Crescent, pp. 201-276.

Geertz, C.: 1973, **The Interpretation of Cultures**, Basic Books, New York.

Geertz, C.: 1980a, Sociosexology, **New York Review of Books**, 26, January 24, pp. 3 - 4.

Geertz, C.: 1980b, **Negara. The Theatre State in Nineteenth Century Bali**, University Press, Princeton.

Ghiselin, M. T.: 1974, **The Economy of Nature and the Evolution of Sex**, University of California Press, Berkeley and Los Angelos.

Ghiselin, M. T.: 1978, 'The Economy of the Body', **American Economic Review**, 68, pp. 233-237.

Giesen, B.: 1980, **Makrosoziologie. Eine evolutionstheoretische Einführung**, Hoffmann & Campe, Hamburg.

Giesen, B.: 1983, 'Moralische Unternehmer und öffentliche Diskussion', **Kölner Zeitschrift für Soziologie und Sozialpsychologie**, 35, pp. 230-254.

Giesen, B.: 1987, 'Beyond Reductionism. Four Models Relating Micro to Macro Level in Sociology', in J. Alexander, B. Giesen, R. Münch, and N. Smelser (eds.): **The Micro-Macro Link**, University of California Press, Berkeley.

Giesen, B., and M. Schmid: 1975, 'System und Evolution. Metatheoretische Vorbemerkungen zu einer soziologischen Evolutionstheorie', **Soziale Welt**, 26, pp. 385- 413.

Giesen, B., and C. Lau: 1981, 'Zur Anwendung darwinistischer Erklärungsstrategien in der Soziologie', **Kölner Zeitschrift für Soziologie und Sozialpsychologie**, 33, pp. 229-256.

Globus, G. G., Maxwell, G., and Savodnik, I. (eds.): 1976, **Consciousness and the Brain**, Plenum Press, New York.

Gluckman, M.: 1977, **Politics, Law, and Ritual in Tribal Society** (4th. ed.), Blackwell, Oxford.

Glück, P., and M. Schmid: 1980, 'Theoretische Falsifikation und Approximation. Zum logischen Verhältnis von Nutzen- und Lerntheorie', **Beiträge zur Philosophie und Methodologie der Erfahrungswissenschaft**, 6, ed. by H. Albert, M. Küttner, und H. Lenk.

Goldschmidt, W.: 1959, **Man's Way: A Preface to the Understanding of Human Society**, Holt, Reinhard and Winston, New York.

Gould, M.: 1976, 'Systems Analysis, Macrosociology and the Generalized Media of Social Action', in J. J. Loubser, R. C. Baum, A. Effrat, and V. M. Lidz (eds.): **Explorations in General Theory in Social Science**, The Free Press, New York, pp. 493-506.

Gould, S. J.: 1978, 'Sociobiology: The Art of Story-Telling', **New Scientist**, 80, pp. 530 - 533.

Gould, S. J.: 1979, 'Episodic Change versus Gradualist Dogma', **Science and Nature**, 2, pp. 5 - 12.

Gould, S. J.: 1982, 'Darwinism and the Expansion of Evolutionary Theory', **Science**, 216, pp. 380-387.

Granovetter, M.: 1979, 'The Idea of 'Advancement' in Theories of Social Evolution', **American Journal of Sociology**, 85, pp. 489-515.

Grove, J. W.: 1980, 'Popper "Demystified": The Curious Ideas of Bloor (and some others) about World 3', **Philosophy of Social Sciences**, 10, pp. 173-180.

Gutmann, W. F., and K. Bonik: 1981, **Kritische Evolutionstheorie: Ein Beitrag zur Überwindung altdarwinistischer Dogmen**, Gerstenberg, Hildesheim.

Haas, E. B.: 1964, **Beyond the Nation State: Functionalism and International Organization**, Stanford University Press, Stanford.

Haas, E. B.: 1970, **The Web of Interdependence**, Prentice Hall, Englewood Cliffs.

Habermas, J.: 1971, 'Vorbereitende Bemerkungen zu einer Theorie der kommunikativen Kompetenz', in J. Habermas and N. Luhmann, **Theorie der Gesellschaft oder Sozialtechnologie. Was leistet die Systemforschung?**, Suhrkamp, Frankfurt, pp. 101-141.

Habermas, J.: 1979, **Communication and the Evolution of Society**, Beacon Press, Boston.

Habermas, J.: 1981, **Theorie kommunikativen Handelns**, 2 vols, Suhrkamp, Frankfurt.

Haken, H.: 1978, **Synergetics: An Introduction**, Springer, Berlin.

Hallowell, A. I.: 1969, 'The Protocultural Foundations of Human Adaptation', in Y. A. Cohen (ed.): **Man in Adaptation - The Biosocial Background**, 3rd printing, Aldine, Chicago, pp. 62-75.

Hamilton, W. D.: 1964a, 'The Genetical Evolution of Social Behaviour. I', **Journal of Theoretical Biology**, 7, pp. 1-16.

Hamilton, W. D.: 1964b, 'The Genetical Evolution of Social Behaviour. II', **Journal of Theoretical Biology**, 7, pp. 17 - 32.

Hammond, M.: 1983, 'Emile Durkheim's The Division of Labor in Society' as a Classic in Human Biosociology', **Journal of Social und Biological Structures**, 6, pp. 123-134.

Hannan, M. and J. Freeman: 1976, 'The Populational Ecology of Organizations', **American Journal of Sociology**, 82, pp. 924-964.

Harris, M.: 1979, **Cultural Materialism: The Struggle for a Science of Culture**, Random House, New York.

Hawthorne, G.: 1976, **Enlightment and Despair. A History of Sociology**, Cambridge University Press, Cambridge et al.

Hayek, F. v.: 1969, **Freiburger Studien. Gesammelte Aufsätze**, Mohr (Siebeck), Tübingen.

Hayek, F. v.: 1973, **Law, Legislation, and Liberty, vol 1. Rules and Order**, Routledge and Kegan Paul, London.

Hayek, F. v.: 1976, **Individualismus und wirtschaftliche Ordnung**, Wolfgang Neugebauer, Salzburg.

Henderson, R. N.: 1972, **The King in Every Man. Evolutionary Trends in Onitsha Ibo Society and Culture**, Yale University Press, New Haven/London.

Hennen, M., and W. U. Prigge: 1977, **Autorität und Herrschaft**, Wissenschaftliche Buchgesellschaft, Darmstadt.

Hirshleifer, J.: 1978, 'Natural Economy versus Political Economy', **Journal of Social and Biological Structures**, 1, pp. 319-337.

Hirschman, A. O.: 1970, **Exit, Voice, and Loyality**, Harvard University Press, Cambridge, Mass.

Ho, M.-W., and P. T. Saunders: 1982, 'The Epigenetic Approach to the Evolution of Organisms - with Notes on its Relevance to Social and Cultural Evolution', in H. C. Plotkin (ed.), **Learning Development, and Culture**, Wiley and Sons, Chichester-New York, pp. 343-361.

Hofstadter, D. R.: 1979, **Gödel, Escher, Bach: An Eternal Golden Braid**, Penguin Books, Harmondsworth.

Hofstadter, D. R., and D. C. Dennett (eds.): 1981, **The Mind's I**, Bantam Books, New York.

Hofstädter, R.: 1969, **Social Darwinism in American Thought**, Braziller, New York.

Hoyer, P.: 1982, **Nature, Human Nature, and Society. Marx, Darwin and the Human Sciences**, Greenwood Press, Westport-London.

Hume, D.: 1978, **Treatise of Human Nature**, P. Nidditch, ed., Oxford University Press, Oxford.

Huntington, P.: 1965, 'Political Development and Political Decay', **World Politics**, 17, pp. 387-430.

Huntington, P.: 1968, **Political Order in Changing Societies**, Yale University Press, New Haven.

Huxley, J.: 1957a, 'Evolution, Cultural and Biological', in J. Huxley, **New Bottles for New Wine: Essays**, Harper & Brothers, New York, pp. 61-92.

Huxley, J.: 1957b, 'Evolutionary Humanism', in J. Huxley, **New Bottles for New Wine: Essays**, pp. 279-312.

Huxley, J.: 1959, 'Man's Place in Nature', in H. V. Hodson (ed.): **The Destiny of Man**, Hodder & Stroughton, London, pp. 13-23.

Huxley, T. H.: 1893, 'Evolution and Ethics, (The Romanes Lecture for 1893)', in **Collected Essays 9, Macmillan**, London.

Jackson, F.: 1980, 'Interactionism Revived?', **Philosophy of the Social Sciences**, 10, pp. 316-323.

Jantsch, E.: 1975, **Design for Evolution: Self-Organization and Planning in the Life of Human Systems**, Braziller, New York.

Kant, I.: 1959, **Foundations of the Metaphysics of Morals**, Trans. L. W. Beck., Bobbs-Merrill, Indianapolis.

Kaplan, M.: 1957, **System and Process in International Politics**, Wiley, New York.

Keohane, R. O., and J. S. Nye: 1977, **Power and Interdependence: World Politics in Transition**, Little, Brown, Boston.

Koch, K.-F.: 1974, **War and Peace in Jalemo. The Management of Conflict in Highland New Guinea**, Harvard University Press, Cambridge, Mass.

Krebs, J. R., and N. B. Davies: 1978, **Behavioral Ecology: An Evolutionary Approach**, Sinauer, Sunderland, Mass.

König, R.: 1973, 'Nachwort', in E. Durkheim, **Der Selbstmord**, Luchterhand, Neuwied, pp. 471-502.

Lakatos, I.: 1968, 'Changes in the Problem of Inductive Logic', in I. Lakatos (ed.), **The Problem of Inductive Logic**, North Holland, Amsterdam, pp. 315-417.

Lakatos, I.: 1970, 'Falsification and the Methodology of Scientific Research Programmes', in I. Lakatos and A. Musgrave (eds.), **Criticism and the Growth of Knowledge**, Cambridge University Press, Cambridge, pp. 91- 195.

Lamnek, S.: 1979, **Theorien abweichenden Verhaltens**, Fink Verlag, München.

Langer, S. K.: 1971, 'The Great Shift: Instinct to Intuition', in J. F. Eisenberg and W. S. Dillon (eds.), **Man and Beast: Comparative Social Behavior**. Smithsonian Annual III, Smithsonian Institution, Washington, pp. 297-313.

Langton, J.: 1979, 'Darwinism and the Behavioural Theory of Sociocultural Evolution: An Analysis', **American Journal of Sociology**, 85, 288-309.

Lasswell, H. D., and A. Kaplan: 1950, **Power and Society. A Framework for Political Inquiry**, Yale University Press, New Haven-London.

Lau, C.: 1981, **Gesellschaftliche Evolution als kollektiver Lernprozeß**, Duncker & Humblot, Berlin.

Le Maho, Y.: 1977, 'The Emperor Penguin: A Strategy to Live and Breed in the Cold', **American Scientist**, 65, pp. 680-693.

Leach, E.: 1976, **Culture and Communication. The Logic by which Symbols are Connected**, University Press, Cambridge.

LeDoux, J. E., Wilson, D. H. and Gazzaniga, M. S.: 1979, 'Beyond Commissurotomy: Clues to Consciousness', in M. S. Gazzaniga (ed.): **Handbook of Behavioral Neurobiology**, vol. 2, Plenum Press, New York, pp. 543-553.

Leinfellner, W.: 1983, 'Das Konzept der Kausalität und der Spiele in der Evolutionstheorie', in K. Lorenz and F. M. Wuketits (eds.), **Die Evolution des Denkens**, Piper, München, pp. 215-261.

Leinfellner, W.: 1984, 'Evolutionary Causality, Theory of Games, and Evolution of Intelligence', in F. M. Wuketits (ed.), **Concepts and Approaches in Evolutionary Epistemology: Towards an Evolutionary Theory of Knowledge**, D. Reidel, Dordrecht-Boston-Lancaster, pp. 233-277.

Lenski, G. E.: 1970, **Human Societies: A Macrolevel Introduction To Sociology**, McGraw Hill, New York.

Levi-Strauss, C.: 1962, **La pensee sauvage**, Plon, Paris.

Levinson, P.: 1982, 'Evolutionary Epistemology Without Limits', **Knowledge: Creation, Diffusion, Utilization** 3, pp. 465-502.

Lewontin, R. C., S. Rose, and L. J. Kamin: 1984, **Not in Our Genes: Biology, Ideology, and Human Nature**, Pantheon Books, New York.

Lidz, Ch. and V. Lidz: 1976, 'Piaget's Psychology of Intelligence and the Theory of Action', in J. Loubser, R. Baum, A. Effrat, and V. Lidz (eds.): **Explorations in General Theory in Social Science: Essays in Honor of Talcott Parsons**, vol. 1, Free Press, New York, pp. 195-230.

Little, R. W.: 1964, 'Buddy Relations and Combat Performance', in M. Janowitz (ed.), **The New Military**, Russell Sage Foundation, New York, pp. 195-225.

Locke, J.: 1975, **An Essay Concerning Human Understanding**, P. H. Nidditch (ed).: Oxford University Press, New York.

Loftus, G.: 1982, 'Open Peer Commentary: Top-down Guidance from a Bottom-up Theory', **The Behavioral and Brain Sciences**, 5, pp. 17-18.

Lomax, A., with N. Berkowitz: 1972, 'The Evolutionary Taxonomy of Culture, **Science**, 177, pp. 228-239.

Lorenz, K.: 1966, **On Aggression**, Methuen, London.

Lorenz, K.: 1973, **Die Rückseite des Spiegels: Versuch einer Naturgeschichte menschlichen Erkennens**, R. Piper, Munich-Zürich. (Engl. translation: Harcourt Brace Jovanovich, New York-London, 1977).

Lorenz, K., and F. M. Wuketits (eds.): 1983, **Die Evolution des Denkens**, Piper, München.

Luckmann, T.: 1980, **Lebenswelt und Gesellschaft**, F. Schoeningh, Paderborn.

Luhmann, N.: 1971a, 'Sinn als Grundbegriff der Soziologie', in J. Habermas and N. Luhmann, **Theorie der Gesellschaft oder Sozialtechnologie. Was leistet die Systemforschung?**, Suhrkamp, Frankfurt, pp. 101-141.

Luhmann, N.: 1971b, 'Systemtheoretische Argumentationen. Eine Entgegnung auf Jürgen Habermas', in J. Habermas and N. Luhmann, **Theorie der Gesellschaft oder Sozialtechnologie. Was leistet die Systemforschung?**, Suhrkamp, Frankfurt, pp. 291-405.

Luhmann, N.: 1972, **Rechtssoziologie**, 2 vols, Rowohlt, Reinbek bei Hamburg.

Luhmann, N.: 1975, **Soziologische Aufklärung 2**, Westdeutscher Verlag, Opladen.

Luhmann, N.: 1976, 'Funktionen und Folgen formaler Organisation', in G. Büschges (ed.): **Organisation und Herrschaft**, Rowohlt, Hamburg, pp. 195-225.

Luhmann, N.: 1984, **Soziale Systeme. Grundriß einer allgemeinen Theorie**, Suhrkamp, Frankfurt.

Luhman, N.: 1985, 'Zum Begriff der sozialen Klasse', in N. Luhman (ed.), **Soziale Differenzierung**, Westdeutscher Verlag, Opladen, pp. 119-162.

Lumsden, Ch. J. and E. O. Wilson: 1981, **Genes, Mind, and Culture**, Harvard University Press, Cambridge, Mass..

Lumsden, Ch. J. and E. O. Wilson: 1982, 'Precis of "Genes, Mind and Culture (With Open Peer Commentary)', **The Behavioral and Brain Sciences**, 5, pp. 1-37.

Lumsden, Ch. J. and E. O. Wilson: 1983, **Promethean Fire**, Harvard University Press, Cambridge, Mass..

MacKay, D. M.: 1978, Selves and Brains, **Neuroscience**, 3, pp. 599-606.

Mackie, J. L.: 1977, **Ethics: Inventing Right and Wrong**, Penguin, Harmondsworth.

Mackie, J. L.: 1978a, 'The Law of the Jungle', **Philosophy**, 53, pp. 553 - 573.

Mackie, J. L.: 1978b, 'Failures in Criticism: Popper and His Commentators', **British Journal for the Philosophy of Science**, 29, pp. 363-387.

Malthus, T. R.: 1826, **An Essay on the Principle of Population**, 6th ed., Johnsen, London.

Mandelbaum, M.: 1971, **History, Man and Reason. A Study in Nineteenth Century Thought**, The John Hopkins University Press, Baltimore-London.

Markl, H.: 1982, 'Evolutionsbiologie des Aggressionsverhaltens', in R. Hilke and W. Kempf (eds.): **Aggression. Naturwissenschaftliche und kulturwissenschaftliche Perspektiven der Aggressionsforschung**, Huber, Bern, pp. 21-44.

Marshall, A.: 1890, **Principles of Economics**, Macmillan, London.

Marx, K.: 1965, **Das Kapital**, vol. 1, Dietz Verlag, Berlin.

Maynard Smith, J.: 1978a, 'The Concepts of Sociobiology', in G. S. Stent (ed.): **Morality as a Biological Phenomenon**, Dahlem Konferenzen, Berlin, pp. 23-33.

Maynard Smith, J.: 1978b, 'Evolution and the Theory of Games', in T. H. Clutton-Brock and P. H. Harvey (eds.), **Readings in Sociobiology**, Freeman, Reading San Francisco, pp. 258-271.

Maynard Smith, J.: 1978c, 'The Evolution of Behavior', **Scientific American**, 239 (3), pp. 176 - 193.

Maynard Smith, J.: 1982, 'Open Peer Commentary: Mind and the Linkage Between Genes and Culture', **The Behavioral and Brain Sciences**, 5, pp. 20-21.

Mayr, E.: 1974, 'Teleological and Teleonomic: A New Analysis', **Boston Studies in the Philosophy of Science**, 14, pp. 91-117.

McNett, C. W., Jr.: 1970a, 'A Settlement Pattern Scale of Cultural Complexity, in R. Naroll and R. Cohen (eds.): **A Handbook of Method in Cultural Anthropology**, Natural History Press, Garden City, pp. 872-886.

McNett, C. W., Jr.: 1970b, 'A Cross-cultural Method for Predicting Nonmaterial Traits in Archeology', **Behavior Science Notes**, 5, pp. 195-212.

McNett, C. W., Jr.: 1973, 'Factor Analysis of a Cross-cultural Sample', **Behavior Science Notes**, 8, pp. 233-257.

Mead, G. H.: 1934, **Mind, Self and Society**, University of Chicago Press, Chicago.

Merton, R. K.: 1964, **Social Theory and Social Structure**, The Free Press, London.

Meyer, P.: 1982, **Soziobiologie und Soziologie: Eine Einführung in die biologischen Voraussetzungen sozialen Handelns**, Luchterhand, Darmstadt-Neuwied.

Michels, R.: 1949 (1911), **Political Parties** (E. Paul and C. Paul trans.), The Free Press, New York.

Mill, J. S.: 1910, **Utilitarianism, Liberty, and Representative Government**, Dent, London.

Miller, J. C.: 1976, **Kings and Kinsmen**, Clarendon Press, Oxford.

Miller, J. G.: 1978, **Living Systems**, McGraw Hill, New York.

Miller, M.: 1986, **Kollektive Lernprozesse: Studien zur Grundlegung einer soziologischen Lerntheorie**, Suhrkamp, Frankfurt.

Mitrany, D.: 1975, **The Functional Theory of Politics**, St. Martin's Press, New York.

Mohr, H.: 1977, **Lectures on Structure and Significance of Science**, Springer, Berlin-Heidelberg-New York.

Moore, G. E.: 1903, **Principia Ethica**, Cambridge University Press, Cambridge.

Moore, B.: 1978, **Injustice. The Social Basis of Obedience and Revolt**, M. E. Sharpe, New York.

Moore, W. E.: 1967, **Strukturwandel der Gesellschaft**, Juventa Verlag München.

Moreno, J.: 1967, **Die Grundlagen der Soziometrie**, Westdeutscher Verlag, Köln.

Mousnier, R.: 1974, **Les institutions de la France sous la monarchie absolue**, Vol. 1. Presses Universitaires de France, Paris.

Murdock, G. P., and C. Provost: 1973, 'Measurement of Cultural Complexity', **Ethnology**, 12, pp. 379-392.

Münch, R.: 1976, **Legitimität und politische Macht**, Westdeutscher Verlag, Opladen.

Münch, R.: 1981, 'Talcott Parsons and the Theory of Action. The Kantian Core', **American Journal of Sociology**, 86, pp. 709-739.

Münch, R.: 1982, **Theorie des Handelns zur Rekonstruction der Beiträge von Talcott Parsons, Emile Durkheim und Max Weber**, Suhrkamp, Frankfurt.

Naroll, R.: 1967, 'Imperial Cycles and World Order', **Peace Research Society Papers**, 7, pp. 83-101.

Naroll, R.: 1970, 'What Have We Learned From Cross-cultural Surveys?', **American Anthropologist**, 72, pp. 1227-1228.

Natsoulas, T.: 1981, 'Basic Problems of Consciousness', **Journal of Personality and Social Psychology**, 41, pp. 132-178.

Nelson, R. J.: 1975, 'Behaviorism, Finite Automata, and Stimulus Theory', **Theory and Decision**, 6, pp. 249-267.

Nelson, R. R., and S. G. Winter: 1982, **An Evolutionary Theory of Economic Change**, The Belknap Press, Cambridge, Mass.-London.

Nisbet, R.: 1969, **Social Change and History. Aspects of Western Theory of Development**, Oxford, University Press New York.

Odum, E. P., and H. T. Odum: 1971 (1953), **Fundamentals of Ecology** (3rd ed.), Saunders, Philadelphia.

Oliver, S. C.: 1974, 'Ecology and Cultural Continuity as Contributing Factors in the Social Organization of the Plains Indians', in Y. A. Cohen (ed.): **Man in Adaptation: The Cultural Present** (2nd ed.), Aldine, Chicago, pp. 302-322.

Opp, K. D.: 1983, **Die Entstehung von Normen**, Mohr (Siebeck), Tübingen.

Ospovat, D.: 1981, **The Development of Darwin's Theory**, Cambridge University Press, Cambridge.

Ott, J. A., G. P. Wagner, and F. M. Wuketits (eds.): 1985, **Evolution, Ordnung und Erkenntnis**, Parey, Berlin-Hamburg.

Otterbein, K.: 1970, **The Evolution of War**, HRAF Press, New Haven.

Ouchi, W.: 1981, **Theory Z**, Addison Wesley, Reading, Mass.

Pantin, C. F. A.: 1968, **The Relations Between the Sciences**, Cambridge Unicersity Press, Cambridge.

Parsons, T.: 1967, **Sociological Theory and Modern Society**, The Free Press, New York-London.

Parsons, T.: 1968, **The Structure of Social Action**, 2 vols, 2nd edition, The Free Press, New York.

Parsons, T.: 1969, **Politics and Social Structure**, The Free Press, New York.

Parsons, T.: 1975, **Gesellschaften. Evolutionäre und komparative Perspektiven**, Suhrkamp, Frankfurt.

Parsons, T.: 1977a, **The Evolution of Societies**, Prentice Hall, Englewood Cliffs.

Parsons, T.: 1977b, **Social Systems and the Evolution of Action Theory**, The Free Press, New York.

Parsons, T.: 1978, **Action Theory and the Human Condition**, The Free Press, New York.

Parsons, T., and N. Smelser: 1956, **Economy and Society**, The Free Press, Glencoe.

Peel, J. D. Y.: 1971, **Herbert Spencer. The Evolution of a Sociologist**, Basic Books, New York.

Perzigian, A. J.: 1981, 'Open Peer Commentary: Genetics, Evolution and Cultural Selection', **The Behavioral and Brain Sciences**, 4, pp. 246-247.

Pianka, E. R.: 1978, **Evolutionary Ecology**, (2nd ed.) Harper and Row, New York.

Pieper, R.: 1979, 'Wissensformen und Rechtfertigungsstrategien', **Soziale Welt**, 30, pp. 50-69.

Pieper, R.: 1987, 'The Self as a Parsite. A Sociological Criticism of Popper's Theory of Evolution, **this volume**.

Pieper, R.: forthcomming, **Die Neue Sozialphysik. Zur Mechanik der Solidarität**.

Pilbeam, D.: 1984, 'The Descent of Hominoids and Hominids', **Scientific American**, 250 (3), pp. 60-69.

Pirages, D. C.: 1978, **The New Context for International Relations: Global Ecopolitics**, Duxbury, North Scituate, Mass.

Platon: 1977, **Der Staat**, Meiner, Hamburg.

Plotkin, H. C.: 1982, 'Evolutionary Epistemology and Evolutionary Theory', in H. C. Plotkin (ed.), **Learning, Development, and Culture: Essays in Evolutionary Epistemology**, Wiley & Sons, Chichester-New York, pp. 3-13.

Plotkin, H. C., and F. J. Odling-Smee: 1981, 'A Multiple-Level Model of Evolution and Its Implications for Sociobiology (with Open Peer Commentary)', **The Behavioral and Brain Sciences**, 4, pp. 225-268.

Poggi, G.: 1972, **Images of Society. Essays in the Sociological Theories of Tocqueville, Marx, and Durkheim**, Stanford University Press, Stanford-London.

Popper, K. R.: 1961, **The Poverty of Historicism**, Routledge & Kegan Paul, London.

Popper, K. R.: 1962, **The Open Society and Its Enemies**, 2 vols., Routledge & Kegan Paul, London.

Popper, K.: 1972, **Objective Knowledge. An Evolutionary Approach**, Oxford University Press, Oxford.

Popper, K.: 1974a, 'Intellectual Autobiography', in P. A. Schilpp, (ed.), **The Philosophy of Karl Popper**, vol. 1, Open Court, La Salle, Ill., pp. 1-181.

Popper, K.: 1974b, 'Scientific Reduction and the Essential Incompleteness of all Science', in F. Ayala, and T. Dobzhansky (eds.): **Studies in the Philosophy of Biology**, Macmillan, London, pp. 259-284.

Popper, K.: 1978a, 'Natural Selection and the Emergence of Mind', **Dialectica**, 32, pp. 339-355.

Popper, K.: 1978b, 'The Propensity Interpretation of Probability', in R. Tuomela (ed.): **Dispositions**, Reidel, Dordrecht and Boston, pp. 247-265.

Popper, K., and J. C. Eccles: 1977, **The Self and Its Brain**, Springer, Berlin and New York.

Powers, W. T.: 1973a, 'Feedback: Beyond Behaviorism', **Science**, 179, pp. 351-356.

Powers, W. T.: 1973b, **Behavior: The Control of Perception**, Aldine, Chicago.

Powers, W. T.: 1978, 'Quantitative Analysis of Purposive Systems: Some Spadework at the Foundations of Scientific Psychology', **Psychological Review**, 85, pp. 417-435.

Prigogine, I.: 1976, 'Order through Fluctuation: Self-Organization and
    Social System', in E. Jantsch and C. H. Waddington (eds): **Evolution
    and Consciousness: Human Systems in Transition**, Addison-Wesley
    Publishing Company, Reading Mass., pp. 93-133.
Prigogine, I., and I. Stengers: 1981, **Dialog mit der Natur: Neue Wege
    naturwissenschaftlichen Denkens**, R. Piper, München-Zürich.
Puccetti, R., and R. W. Dykes: 1978, 'Sensory Cortex and the Mind-Brain
    Problem (with Open Peer Commentary)', **The Behavioral and Brain
    Sciences**, 3, pp. 337-375.
Putnam, H.: 1975, **Mind, Language and Reality**, Cambridge University Press,
    Cambridge, Mass.
Pylyshyn, Z. W.: 1980, 'Computation and Cognition: Issues in the
    Foundation of Cognitive Science', **The Behavioral and Brain Sciences**,
    3, pp. 111-169.
Quinton, A.: 1973, 'Sir Karl Popper, Knowledge as an Institution',
    Encounter, 16, pp. 33-37.
Radnitzky, G.: 1980, 'From Justifying a Theory to Comparing Theories and
    Selecting Questions', **Revue Internationale de Philosophie**, 34, pp.
    179-227.
Radnitzky, G., and G. Andersson (eds.): 1978, **Progress and Rationality in
    Science**, D. Reidel, Dordrecht-Boston-London.
Rappaport, R. A.: 1979, **Ecology, Meaning, and Religion**, North Atlantic
    Books, Richmond, Cal.
Rawls, J.: 1971, **A Theory of Justice**, Harvard University Press, Cambridge,
    Mass.
Renfrew, C.: 1984, **Approaches to Social Archeology**, Edinburgh
    University Press, Edinburgh.
Riedl, R.: 1975, **Die Ordnung des Lebendigen: Systembedingungen der
    Evolution**, Parey, Berlin-Hamburg. (English translation: Wiley & Sons,
    New York, 1979).
Riedl, R.: 1977, 'A Systems-Analytical Approach to Macro-Evolutionary
    Phenomena', **The Quarterly Review of Biology**, 52, pp. 351-370.
Riedl, R.: 1980, **Biologie der Erkenntnis: Die stammesgeschichtlichen
    Grundlagen der Vernunft**, Parey, Berlin-Hamburg.
Riedl, R.: 1984: 'Evolution and Evolutionary Knowledge: On the
    Correspondence between Cognitive Order and Nature', in F. M. Wuketits

(ed.): **Concepts and Approaches in Evolutionary Epistemology. Towards an Evolutionary Theory of Knowledge**, Reidel, Dordrecht-Boston-Lancaster pp. 35-50.

Riedl, R., and F.M. Wuketits (eds.) 1987: **Die Evolutionäre Erkenntnistheorie**, Parey, Berlin-Hamburg.

Ritzer, G.: 1975, **Sociology. A Multiple Paradigm Science**, Allyn and Bacon, Boston.

Roberts, S.: 1979, **Order and Dispute. An Introduction to Legal Anthropology**, Penguin Books, Harmondsworth.

Rorty, A. O.: 1976, 'Introduction', in A. O. Rorty (ed.): **The Identities of Persons**, University of California Press, Berkeley, pp. 1-15.

Rosenberg, A.: 1985, **The Structure of Biological Science**, Cambridge University Press, Cambridge-London-New York.

Rossiter, C. L.: 1948, **Constitutional Dictatorship: Crisis Government in the Modern Democracies**, Princeton University Press, Princeton.

Ruse, M.: 1979, **The Darwinian Revolution: Science Red in Tooth and Claw**, University of Chicago Press, Chicago.

Ruse, M.: 1980, 'Charles Darwin and Group Selection', **Annals of Science**, 37, pp. 615 - 630.

Ruse, M.: 1982, **Darwinism Defended. A Guide to the Evolution Controversies**, Addison-Wesley, London et al.

Ruse, M.: 1984a, **Sociobiology: Sense or Nonsense**, 2nd ed., Reidel, Dordrecht.

Ruse, M.: 1984b, 'The Morality of the Gene', **The Monist**, 67, pp. 167-199.

Ruse, M.: 1984c, 'The Morality of the Gene', **The Monist**, 67, pp. 167-199.

Ruse, M.: 1986, **Taking Darwin Seriously: A Naturalistic Approach to Philosophy**, Blackwells, Palo Alto.

Ruse, M., and E. O. Wilson: 1986, 'Ethics as Applied Science' **Philosophy**, forthcoming.

Russell, B.: 1968, **Unpopular Essays**, Allen & Unwin, London.

Russett, B. M.; 1967, 'The Ecology of Future International Politics', **International Studies Quarterly**, 2, pp. 12-31.

Russett, C. E.: 1976, **Darwin in American: The Intellectual Response. 1865 - 1912**, Freeman, San Francisco.

Ryle, G.: 1949, **The Concept of Mind**, Hutchison, London.

Sahlins, M. D.: 1960, 'Evolution: Specific and General', in M. D. Sahlins and E. R. Service (eds.): **Evolution and Culture**, University of Michigan Press, Ann Arbor.

Sahlins, M. D.: 1962, 'Poor Man, Rich Man, Big Man, Chief. Political Types in Melanesia and Polynesia', **Comparative Studies in Society and History**, 5, 285-303.

Sahlins, M. D.: 1976, **The Use and Abuse of Biology**, University of Michgan Press, Ann Arbor.

Sahlins, M.: 1981, **Historical Mataphors and Mythical Realitiesf. Structure in the Early History of the Sandwich Islands Kingdom.** University of Michigan Press, Ann Arbor.

Sahlins, M. D., and E. Service: 1960, **Evolution and Culture**, University of Michigan Press, Ann Arbor.

Saxe, A. A.: 1977, 'On the Origins of Evolutionary Processes: State Formation in the Sandwich Islands', in J. N. Hill (ed.), **Explanation of Prehistoric Change**, University of New Mexico Press, Albuquerque, pp. 105-154.

Schilpp, P. A. (ed.): 1974, **The Philosophy of Karl Popper**, 2 vols., Open Court, La Salle, Ill.

Schlesinger, A. M.: 1939, 'Tides of American Politics', **Yale Review**, 29, pp. 217-230.

Schmid, M.: 1981, 'Struktur und Selektion. Emile Durkheim und Max Weber als Theoretiker struktureller Selektion', **Zeitschrift für Soziologie**, 10, pp. 17-37.

Schmid, M.: 1982a, 'Habermas' Theory of Social Evolution', in J. B. Thompson and D. Held (eds.), **Habermas - Critical Debates**, MacMillan, London, pp. 163-180.

Schmid, M.: 1982b, 'Werte und soziale Integration', in J. Becker, I. Lichtenstein-Rother and H. Stopp (eds), **Wertepluralismus und Wertewandel heute**, Ernst Vögel, München, pp. 19-30.

Schmid, M.: 1982c, **Theorie sozialen Wandels**, Westdeutscher Verlag, Opladen.

Schmid, M.: 1983, **Gleichgewicht und Evolution. Bemerkungen zur Theorie sozialen Wandels von Talcott Parsons**, Typoscript, Augsburg.

Schmid, M.: 1987a, 'Collective Action and the Selection of Rules', **this volume.**

Schmid, M.: 1987b, 'Autopoiesis und soziales System. Eine Standortbestimmung', in H. Haferkamp and M. Schmid (eds.), **Sinn, Kommunikation und soziale Differenzierung. Beiträge zu Luhmanns Theorie sozialer Systeme**, Suhrkamp, Frankfurt, pp. 25 - 50.

Schneirla, T. C.: 1951, 'The 'Levels' Concept in the Study of Social Organization in Animals', in J. Rohrer and M. Sherif (eds.): **Social Psychology at the Crossroads**, Harper, New York, pp. 83-120.

Schotter, A.: 1981, **The Economic Theory of Social Institutions**, Cambridge University Press, Cambridge.

Schwartz, B.: 1981, **Vertical Classification. A Study in Structuralism and in the Sociology of Knowledge**, University of Chicago Press, Chicageo.

Schwartz, R. D., and J. C. Miller: 1970, 'Legal Evolution and Societal Complexity', in S. N. Eisenstadt (ed.): **Readings on Social Evolution and Development**, Pergamon, London, pp. 155-172.

Searle, J. R.: 1969, **Speech Acts**, Cambridge University Press, Cambridge, Mass.

Searle, J. R.: 1978, Sociobiology and the Explanation of Behavior, in M. S. Gregory, A. Silvers and D. Sutch (eds.): **Sociobiology and Human Nature**, Jossey-Bass, San Francisco, pp. 164-183.

Service, E. R.: 1962, **Primitive Social Organization: An Evolutionary Perspective**, Random House, New York.

Service, E. R.: 1975, **Origins of the State and Civilisation. The Process of Cultural Evolution**, Norton, New York.

Settle, T.: 1981, 'Letter to Mario: The Self and its Mind', in J. Agassi, and R. S. Cohen (eds.): **Scientific Philosophy Today**, Reidel, Dordrecht, pp. 357-379.

Sewell, J. P.: 1966, **Functionalism and World Politics**, Princeton University Press, Princeton.

Shepher, J.: 1971, 'Mate Selection Among Second Generation Kibbutz Adolescence, Incest Avoidence and Negative Imprinting', **Archives of Sexual Behaviour**, 1, pp. 293-307.

Simon, H. A.: 1957, **Models of Man**, Wiley, New York.

Simon, H. A.: 1969, **The Science of the Artificial**, The M.I.T. Press, Cambridge, Mass.

Simpson, G. G.: 1974, 'The Concept of Progress in Organic Evolution', **Social Research**, 41, pp. 28-51.

Singer, P.: 1972, 'Famine, Affluence, and Morality', **Philosophy and Public Affairs**, 1 (3).

Singer, P.: 1981, **The Expanding Circle: Ethics and Sociobiology**, Farrar, Straus, and Giroux, New York.

Skalnik, P.: 1978, 'The Early State as a Process', in H. J. M. Claessen and P. Skalnik (eds.), **The Early State**, Mouton, The Hague/Paris/New York, pp. 597-635.

Smith, A. D.: 1973, **The Concept of Social Change. A Critique of the Functionalist Theory of Social Change**, Routledge and Kegan Paul, London - Boston.

Smith, A.: 1976 (1776), **An Inquiry into the Nature and Causes of the Wealth of Nations**, University of Chicago Press, Chicago.

Smith, A.: 1977^2, **Theorie der Gefühle**, Meiner, Hamburg.

Spencer, H.: 1857 (1891), 'Progress: Its Law and Cause', **Westminster Review**. Reprinted in H. Spencer, **Essays: Scientific, Political, and Speculative**, Williams and Norgate, London, pp. 8-62.

Spencer, H.: 1862, **First Principles**, Watts, London.

Spencer, H.: 1892, **Principles of Ethics**, Williams and Norgate, London.

Spencer, H.: 1897, **The Principles of Sociology**, vol. 3, Appelton, New York.

Sperry, R. W.: 1965, 'Mind, Brain, and Humanist Values', in J. R. Platt (ed.): **New Views on the Nature of Man**, Chicago University Press, Chicago, pp. 71-92.

Sperry, R. W.: 1976, 'Mental Phenomena as Causal Determinants Brain Function', in G. G. Globus, G. Maxwell and J. Savodnik (eds.), **Consciousness and the Brain**, Plenum Press, New York, pp. 163-177.

Sperry, R. W.: 1980, 'Mind-Brain Interaction: Mentalism, Yes; Dualism, No', **Neuroscience**, 5, pp. 195-206.

Sprout, H., and M. Sprout: 1965, **The Ecological Perspective on Human Affairs, with Special Referenve to International Politics**, University of Princeton Press, Princeton.

Sprout, H., and M. Sprout: 1968, **An Ecological Paradigm for the Study of International Politics**, Center of International Studies, Princeton.

Stebbins, G. L.: 1967, 'Pitfalls and Guideposts in Comparing Organic and Social Evolution', in W. E. Moore and R. M. Cook (ed.): **Readings on Social Change**, Prentice Hall, Englewood Cliffs, pp. 223-234.

Steinbrunner, J. D.: 1974, **The Cybernetic Theory of Decision**, University of Princeton Press, Princeton.

Sumner, W. G.: 1918, **The Forgotten Man and Other Essays**, ed. A. G. Keller, Yale University Press, New Haven.

Szentagotai, J.: 1978, 'Open Peer Commentary: A False Alternative', **The Behavioral and Brain Sciences**, 3, pp. 367-368.

Taagepera, R.: 1968, 'Growth Curves of Empires', **General Systems**, 13, pp. 171-175.

Taagepera, R.: 1978a, 'Size and Duration of Empires: Systematics of Size', **Social Science Research**, 7, pp. 108-127.

Taagepera, R.: 1978b, 'Size and Duration of Empires: Growth-Decline Curves, 3000 to 600 B. C', **Social Science Research**, 7, pp. 180-196

Taagepera, R.: 1979, 'Size and Duration of Empires: Growth-Decline Curves, 600 B. C. to 600 A. D.', **Social Science History**, 3, pp. 115-138.

Taagepera, T.: 1981, 'Growth-Decline Curves of Empires: Some Regularities', **Paper** prepared for the 22nd Annual Convention of International Studies Association, Philadelphia, Pa., March 18-21.

Tennant, N.: 1983, 'Evolutionary **vs.** Evolved Ethics', **Philosophy**, 58, pp. 289-302.

Thorndike, E. L.: 1911, **Animal Intelligence**, Macmillan, New York.

Thorndike, E. L.: 1925, **Educational Psychology: The Psychology of Learning**, vol. 2, Teachers College, New York.

Topitsch, E.: 1978, **Erkenntnis und Illusion: Die Grundstrukturen unserer Weltauffassung**, Hoffmann und Campe, Hamburg.

Touraine, A.: 1973, **Production de la societe**, Seuil, Paris.

Trigg, R.: 1982, **The Shaping of Man**, Blackwells, Oxford.

Trivers, R. L.: 1971, 'The Evolution of Reciprocal Altruism', **Quarterly Review of Biology**, 46, pp. 35 - 57.

Turner, J. H.: 1985, **Herbert Spencer. A Renewed Appreciation**, Sage, Beverley Hills-London-New Delhi.

Turner, V.: 1969, **The Ritual Process. Structure and Anti-Structure**, Cornell University Press, Ithaca.

Ullmann-Margalit, E.: 1977, **The Emergence of Norms**, Clarendon Press, Oxford.

Valjavec, F.: 1985, **Identite sociale et evolution. Elements d'une theorie des processus adaptifs**, Peter Lang, Frankfurt-Bern-New York.

Van Parijs, P.: 1981, **Evolutionary Explanation in the Social Sciences**. An Emerging Paradigm, Tavistock, London-New York.

Vanberg, V.: 1982, **Markt und Organisation**, Mohr (Siebeck), Tübingen.

Vanberg, V.: 1983, 'Der individualistische Ansatz zu einer Theorie der Entstehung und Entwicklung von Institutionen', **Jahrbuch für Neue Politische Ökonomie**, 2, Mohr (Siebeck), Tübingen, pp. 50-69.

Vogel, E. F.: 1980, **Japan as Number One: Lessons for America**, Harper and Row, New York.

Vollmer, G.: 1984, 'Mesocosm and Objective Knowledge: On Problems Solved by Evolutionary Epistemology', in F. M. Wuketits (ed.), **Concepts and Approaches in Evolutionary Epistemology. Towards an Evolutionary Theory of Knowledge**, D. Reidel, Dordrecht-Boston-Lancaster, pp. 69-121.

Waddington, C. H.: 1975, **The Evolution of an Evolutionist**, University of Edinburgh Press, Edinburgh.

Wagner, G. P.: 1981, 'Feedback Selection and the Evolution of Modifiers', **Acta Biotheoretica**, 30, pp. 79-102.

Wagner, G. P.: 1983, 'On the Necessity of a Systems Theory of Evolution and its Population-Genetic Foundation', **Acta Biotheoretica**, 32, pp. 223-226.

Wallace, A. R.: 1870, **Contributions to the Theory of Natural Selection**, Macmillan, London.

Waltz, K. N.: 1975, 'Theory of International Politics', in F. I. Greenstein and N. W. Polsby (eds.): **Handbook of Political Science** vol. 8, Addison Wesley, Reading, Mass.

Waltz, K. N.: 1979, **Theory of International Politics**, Addison-Wesley, Reading, Mass.

Washburn, S. L.: 1978, 'The Evolution of Man', **Scientific American**, 239 (3), pp. 146-154.

Weber, M.: 1922, **Gesammelte Aufsätze zur Wissenschaftslehre**, J. C. B. Mohr, Tübingen, pp. 291-359.

Weber, M.: 1964, **Wirtschaft und Gesellschaft**, 2 vols., Kiepenheuer & Witsch, Köln-Berlin.

Weber, M.: 1965, **Die protestantische Ethik. Eine Aufsatzsammlung**, Siebenstern Taschenbuch Verlag, München-Hamburg.

Werth, R.: 1980, 'Review (The Self and its Brain)', **Erkenntnis**, 15, pp. 409-416.

Wesson, R. G.: 1967, **The Imperial Order**, University of California Press, Berkeley and Los Angelos.

White, E.: 1983, 'Brain Science and the Emergence of Neuropolitics', **Politics and the Life Sciences**, 1, pp. 23-25.

Whyte, W. F.: 1949, 'The Social Structure of the Restaurant', **American Journal of Sociology**, 54, pp. 302-310.

Wiener, N.: 1948, **Cybernetics**, Wiley, New York.

Willecke, F. U.: 1961, **Die Entwicklung der Markttheorie. Von der Scholastik bis zur Klassik**, Mohr (Siebeck), Tübingen.

Williams, G. C.: 1966, **Adaptation and Natural Selection: A Critique of Some Current Evolutionary Thought**, Princeton University Press, Princeton.

Wilson, E. O.: 1975, **Sociobiology: The New Synthesis**, Belknap, Cambridge, Mass.

Wilson, E. O.: 1978, **On Human Nature**, Harvard University Press, Cambridge, Mass.

Wimberly, H.: 1973, 'Legal Evolution: One Further Step', **American Journal of Sociology**, 79, pp. 78-83.

Wimsatt, W. C.: 1976, 'Reductionism, Levels of Organization, and the Mind-Body Problem', in G. G. Globus, G. Maxwell and J. Savodnik (eds.), **Consciousness and the Brain**, Plenum Press, New York, pp. 205-267.

Winch, P.: 1958, **The Idea of a Social Science**, Routledge and Kegan Paul, London.

Wirsing, R.: 1973, 'Political Power and Information: A Cross-Cultural Study', **American Anthropologist**, 75, pp. 153-170.

Wolff, K. H. (ed.): 1960, **Emile Durkheim: 1858-1917**, Ohio State University Press, Columbus, Ohio.

Woodcock, A., and M. Davis: 1980, **Catastrophe Theory**, Penguin Books, Harmondsworth.

Wright, H. T.: 1977, 'Toward an Explanation of the Origin of the State', in J. N. Hill (ed.), **Explanation of Prehistoric Change**, University of New Mexico Press, Albuquerque, pp. 215-230.

Wrong, D. H.: 1979, **Power. Its Forms, Bases and Uses**, Basil Blackwell, Oxford.

Wuketits, F. M.: 1979, 'Gesetz und Freiheit in der Evolution der Organismen', **Umschau in Wissenschaft und Technik**, 79, pp. 268-275.

Wuketits, F. M.: 1980, 'On the Notion of Teleology in Contemporary Life Sciences', **Dialectica** 34, pp. 277-290.

Wuketits, F. M.: 1981, **Biologie und Kausalität: Biologische Ansätze zur Kausalität, Determination und Freiheit,** Parey, Berlin-Hamburg.

Wuketits, F. M.: 1982a, 'Systems Research: The Search for Isomorphism', in **Progress in Cybernetics and Systems Research,** vol. 11, Hemisphere Publishing Corporation, Washington.

Wuketits, F. M.: 1982b, **Grundriss der Evolutionstheorie,** Wissenschaftliche Buchgesellschaft, Darmstadt.

Wuketits, F. M.: 1983, 'Evolutionary Epistemology, Objective Knowledge, and Rationality: The Evolutionary Approach in Man's Search for Himself', in **Proceedings of the Eleventh International Conference on the Unity of the Sciences,** vol. 2, International Cultural Foundation Press, New York, pp. 881-899.

Wuketits, F. M. (ed.), 1984a, **Concepts and Approaches in Evolutionary Epistemology: Towards an Evolutionary Theory of Knowledge,** D. Reidel, Dordrecht-Boston Lancaster.

Wuketits, F. M.: 1984b, 'Evolutionary Epistemology: A Challenge to Science and Philosophy', in F. M. Wuketits (ed.), **Concepts and Approaches in Evolutionary Epistemology. Towards an Evolutionary Theory of Knowledge,** D. Reidel, Dordrecht-Boston-Lancaster, pp. 1-33.

Wuketits, F. M.: 1984c, 'Evolutionary Epistemology: A New Copernican Revolution?', in F. M. Wuketits (ed.), **Concepts and Approaches in Evolutionary Epistemology. Towards an Evolutionary Theory of Knowledge,** D. Reidel, Dordrecht-Boston-Lancaster, pp. 279-284.

Wuketits, F. M.: 1985, **Zustand und Bewußtsein: Leben als biophilosophische Synthese,** Hoffmann und Campe, Hamburg.

Wuketits, F. M.: 1987, 'Evolution, Causality, and Freedom. The Open Society from a Biological Point of View', **this volume.**

Young, R. M.: 1969, 'Malthus and the Evolutionists: The Common Context of Biological and Social Theory', **Past and Present,** 43, pp. 109 - 145.

Zeeman, E. C.: 1977, **Catastrophe Theory. Selected Papers 1972 - 1977,** Benjamin, Reading.

Zeleny, M. (ed.): 1981, **Autopoiesis: A Theory of Living Organization,** Elsevier North Holland, New York.

Zimmer, D.: 1981, **Im Dickicht der Gefühle**, Zeit-Magazin, No. 11, pp. 16-
    24.

Zinnes, D. A.: 1975, 'Research Frontiers in the Study of International
    Politics', in F. I. Greenstein and N. W. Polsby (eds.): **Handbook of
    Political Science**, vol. 8, Addison-Wesley, Reading, Mass., pp. 87-198.

INDEX OF NAMES

Abercrombie, N. 81
Adams, R. N. 14
Agassi, J. 10,11,12,13
Albert, H. 97
Aldrich, H. E. 90
Alexander, R. D. 32
Alland, A. 15
Almond, G. A. and J. S.
  Coleman 157
Althusser, L. and E. Balibar
  209,224
Anderson, A. R. and O. K. Moore
  218,220
Andreski, S. 160
Aristoteles, 141,143
Ashby, W. R. 138
Axelrod, R. 164
Axelrod, R. and W. D. Hamilton
  164

Barash, D. 9,16,20,31
Bartley, W. W. 64,66,76
Bateson, G. 4,5
Baum, R. C. 177
Beloff, J. 223
Berger, P. L. and T. Luckmann
  71
Bergson, H. 52
Bertalanffy, L. von 4,55,56,76,
  138
Blau, P. M. 87
Bloch, M. 125
Block, N. 223
Blute, M. 131,135
Boehm, C. 131
Bohannan, P. 8,210
Bohm, D. 208,224
Bonik, K. 52
Boudon, R. 99
Boulding, K. 173,194
Bourdieu, P. 90,188
Bridgeman, B. 224
Brown, R. and R. J. Herrnstein
  130
Buckley, W. 138
Bunge, M. 67,203,204,223
Burrow, J. W. 99

Campbell, D. T. 6,64,76,108,131,
  194,203
Candland, D. K. 130
Carneiro, R. L. 99,148,151,167
Carr, B. 223
Chase, R. 131
Chomsky, N. 206
Claessen, D. and P. Skalinik
  113,125
Claessen, H. J. M. 125
Claessens, D. 1,14
Clark, B. 131
Cloak, F. T. Jr. 224
Cohen, R. 125
Cohen, R. and E. R. Service 111
Compton, A. H. 199,200,206,222
Comte, A. 2,4,21
Corning, P. A. V,60,94,98,127,
  131,132,133,136,138,140,142,
  144,148,152,154,156,159,161,
  163,164,168
Coulam, R. F. 138
Count, E. W. 11,13
Currie, G. 223,224

Darwin, C. II,3,4,23-30,37,38,
  40,43-45,49,51-53,58,81,82,
  99,128-130,163,166,167,170,
  199,202,223
Dawkins, R. 7,29,196
Dennett, D. C. 223,224
Descartes, R. 199,202,203
Deutsch, K. W. 138,160
Dewan, E. M. 224
Dobzhansky, T. 6
Dougherty, J. E. and R. L.
  Pfaltzgraff, Jr. 161
Douglas, M. 21,
Dumont, L. 105,107
Durham, W. H. 7,131
Durkheim, E. 1,2,85,94,99,196,
  197,209-213,217,219,224
Durkheim, E. and M. Mauss 224

Easton, D. 138
Eccles, J. C. 21,195,201,203,
  206,219,224

Eckberg, D. L. 215
Economos, J. 203
Eddington, A. S. 10
Eder, к. V.91,98,99,101,102,
   113,121
Eigen, M. 56
Eigen, M. and ιP. Schuster 56,57
Eigen, M. and R. Winkler 3, 4
Einstein, A. 127,215
Eisenstadt, S. N. 156
Eisenstadt, S. N., and M.
   Curelaru 85
Ekman, P. 17,20
Eldredge, N. 76
Elster, J. 91
Erikson, E. H. 5

Fairservis, W. A. 102,122
Fararo, T. 97
Flannery, K. 152
Flew, A.G.N. 25,28
Ford, H. 132
Francis, E. K. 81
Fried, M. H. 111,113
Friedmann, J. and M. J.
   Rowlands 112

Galilei, G. 53
Geertz, C. 32,106,109
Ghiselin, M. T. 164
Giesen, B. V,83,87,95,99,171,
   172,221
Giesen, B. and C. Lau 83,98,99,
   173
Giesen, B. and M. Schmid 83
Globus, G. G., Maxwell, G. and
   Savodnik, I. 223
Gluckman, M. 110,121
Glück, P. and M. Schmid 99
Goldschmidt, W. 152
Gould, M. 177
Gould, S. J. 33,52,76
Granovetter, M. 81,173
Grove, J. W. 223,224
Gutmann, W. F. and K. Bonik 52

Haas, E. B. 166
Habermas, J. 95,99,102,103,107,
   172,178,194,221

Haeckel, E. 51
Haken, H. 209,210,224
Hallowell, A. I. 15
Hamilton, W. D. 30
Hammond, M. 1,2,11,19,22
Hannan, M. and J. Freeman 90
Harris, M. 108
Hawthorne, G. 80
Hayek, F. v. 82,92
Hegel, G. W. F. 198
Henderson, R. N. 114
Hennen, M. and W. U. Prigge 91
Hirshleifer, J. 174
Hirshman, A. O. 95
Hitler, A. 144
Ho, M. W. and P. T. Saunders
   101
Hobbes, T. 120
Hofstadter, D. R. 223
Hofstadter, D. R. and D. C.
   Dennett 223
Hofstädter, R. 81
Hoyer, P. 81
Hume, D. 28,37-39
Huntington, P. 158,159
Husserl, E. 14
Huxley, J. 50,51,73
Huxley, T. H. 27,50

Jackson, F. 223
Jantsch, E. 61,68

Kant, I. 38,42,213
Kaplan, A. 91
Kaplan, M. 162
Keohane, R. O., and J. S. Nye
   166
Koch, K.-F. 121
Krebs, J. R. and N. B. Davies
   163
König, R. 224

Lakatos, I. 85,190
Lamnek, S. 90
Langer, S. K. 14
Langton, J. 102
Lasswell, H. D. and A. Kaplan
   91
Lau, C. 83

Le Maho, Y.   136
Leach, E.   109
LeDoux, J. E.   224
Leibniz, G. W.   49
Leinfellner, W.   5,6,8,9,21,56,
   58,59,71
Lenski, G. E.   152-154
Levi-Strauss, C.   104,109
Levinson, P.   65
Lewontin, R. C.   31,33,77
Lidz, Ch. and V. Lidz   213
Lincoln, A.   158
Little, R. W.   20
Locke, J.   34
Loftus, G.   217
Lomax, A. with N. Berkowitz
   150,151
Lorenz, K.   29,55,64,66,76
Luckmann, T.   1,14
Luhmann, N.   14,90,98,107,176
Lumsden, Ch.   33,34
Lumsden, Ch. and E. O. Wilson
   217

Machiavelli, N.   143
MacKay, D. M.   203,204,223
Mackie, J. L.   38,41,223,224
Malthus, T. R.   24-26,82
Mandelbaum, M.   2,21,80
Mandler, G.   224
Mannheim, K.   198
Markl, H.   20
Marshall, A.   163
Marx, K.   33,81,99,103.104,107,
   121,152
Mauss, M.   8
Maxwell, B.   224
Maynard Smith, J.   6-9,21,30,
   217
Mayr, E.   55
McNett, C. W., Jr.   150
Mead, G. H.   218,219
Merton, R. K.   87,99,155
Meyer, P.   IV,1,8,51,77
Michels, R.   142,143
Mill, J. S.   43
Miller, J. C.   114,115
Miller, J. G.   138,140,141
Miller, M.   104,110
Mitrany, D.   166
Mohr, H.   55,64

Monod, J.   166
Moore, B.   94
Moore, G. E.   38
Moore, W. E.   99
Moreno, J.   19
Mousnier, R.   107
Murdock, G. P. and C. Provost,
   148
Murray jr., J. J.   131
Münch, R.   172,177

Naroll, R.   151,152,166.167
Natsoulas, T.   215,223
Nelson, R. J.   205
Nelson, R. R. and S. G. Winter,
   90
Newton, I.   53,80,99,128
Nisbet, R.   80, 173

Odum, E. P. and H. T. Odum
   163
Oliver, S. C.   154
Opp, K. D.   91
Ospovat, D.   25
Ott, J. A.   52
Otterbein, K.   137,151
Ouchi, W.   137

Pantin, C. F. A.   133
Pareto, V.   85
Parsons, T.   15,81,85-87,91,99,
   176,184,194,213,214
Parsons, T. and N. J. Smelser,
   176
Peel, J. D. Y.   80,99
Perzigian, A. J.   224
Pianka, E. R.   163
Pieper, R.   VI,99,195,222
Pilbeam, D.   63
Pirages, D. C.   163
Pittendrigh, C.   76
Platon   44,59,60,68,99,143,198,
   221
Platt, J.   168
Plotkin, H. C.   64
Plotkin, H. C. and F. J. Odling-
   Smee, 217
Poggi, G.   79

Popper, K. R.   VI,6,49,50,59,60,
    62,64,66,67,72,73,75,76,173,
    194ff.
Popper, K. R. and J. C. Eccles,
    21
Powers, W. T.   138
Prigogine, I.`` 3,60,61,209,224
Prigogine, I. and I. Stengers,
    3-5,61
Puccetti, R. and R. W. Dykes,
    224
Putnam, H.   206,223
Pylyshyn, Z. W.   206,217

Quetelet, L. A. J.   212,224
Quinton, A.   224

Radnitzky, G. and G. Andersson,
    72
Rappaport, R. A.   106
Rawls, J.   35-37
Renfrew, C.   97
Riedl, R.   4,6,13,14,51-53,60,
    62,64,67,72,76
Ritzer, G.   85
Roberts, S.   90
Rorty, A. O.   223,224
Rosenberg, A.   53,54
Rossiter, C. L.   158
Röpke, J.   194
Ruse, M.   IV,23,24,26,28,29,31,
    33,34,38,39,72,77,81,99
Russell, B.   72
Russett, B. M.   163
Russett, C. E.   27
Ryle, G.   222

Sahlins, M. D.   32,106,114,129,
    172
Sahlins, M. D. and E. R. Service,
    173
Saxe, A. A.   125
Schäffle, A.   197
Schlesinger, A. M.   160
Schmid, M.   V,79,84,87,96,97,99,
    103,111,172,173
Schneirla, T. C.   22
Schotter, A.   91
Schrödinger, E.   6

Schütz, A.   14
Schuster, M.   56,57
Schwartz, B.   109
Schwartz, R. D. and J. C. Miller,
    151
Searle, J. R.   8,216
Service, E. R.   111,113,152
Settle, T.   223,224
Sewell, J. P.   166
Shakespeare, W.   160
Shearmur, J.   207
Shepher, J.   16,17,19
Simon, H. A.   215
Simpson, G. G.   3,21
Singer, P.   37,42
Skalnik, P.   125
Smith, A.   82,92,94,133,134,152
Smith, A. D.   173,174
Spencer, H.   I,26,27,37,40,44,51,
    81,94,99,129,164,167,197,209,
    212
Sperry, R. W.   21,203,204,206,
    209,223,224
Sprout, H. and M. Sprout,   164
Stanley, S. M.   76
Stebbins, G. L.   82
Steinbrunner, J. D.   138
Stengers, J.   3-5,61
Strawson, P. F.   214
Sumner, W. G.   27
Szentagotai, J.   224

Taagepera, R.   156
Taagepera, T.   156
Tennant, N.   72
Thorndike, E. L.   130,132
Topitsch, E.   72
Toulmin, S.   194
Touraine, A.   101
Thrasymochus   44,45
Trigg, R.   37
Trivers, R. L.   7,31,32
Turner, J. H.   80,99
Turner, V.   105

Ullmann-Margalit, E.   91-93,99

Valjavec, F.   99
Van Parijs, P.   98

Vanberg, V.   87
Vogel, E. F.   137
Vollmer, G.   64,76

Waddington, C. H.   90
Wagner, G. P.   52,76
Wallace, A. R.   29
Waltz, K. N.   159-162
Washburn, S. L.   63
Weber, M.   1,85,87,91,94,95,99
Wedgwood, J.   26
Weiss, P. A.   76
Werth, R.   223
Wesson, R. G.   156
White, E.   8
Whitehead, A. N.   220
Whyte, W. F.   134
Wiener, N.   138
Willecke, F. U.   92
Williams, G. C.   27
Wilson, E. O.   7,8,16,29,31,33,
    34,39,77,164,194
Wimberly, H.   151
Wimsatt, W. C.   202
Winch, P.   90
Wirsing, R.   148
Woodcock, A. and M. Davis,   97
Wright, H. T.   125
Wrong, D. H.   91
Wuketits, F. M.   IV,2,4,49,52,53,
    55,56,59,60,61,64,99

Young, R. M.   24

Zeeman, E. C.   97
Zeleny, M.   223
Zimmer, D.   22
Zinnes, D. A.   161

INDEX OF SUBJECTS

Action 1-21,23,31ff,50ff,63,68,
    79-99,108ff. See adaption
Adaptability. See adaption
Adaption 16,26,39,45,51,68,88,
    171,217
Adaptive simplification 155
Affectivity 11f,13f,17ff,210,211
Altruism 7,29ff,35
Anthropology 1,8,148
Authority 31f,93,95,106,116ff,184
Autopoiesis 99,125,223

Behaviour. See action
Behaviourism 8,130
Biogram 8,11,14,17ff
Biosociology 12ff,15ff,51
Brain 64f,204,214,224

Capitalism 26,27,32
Causality III,10ff,49-77,81,130ff,
    196,200,203,206,208,214,217
Code 175,187,193
Co-evolution 218
Communication 94f,104,110f,122,137,
    140,144,176,219
Community 179f,184,188
Competition 16,58,87,92ff,95,137,
    163,185
Complexity 54,129,147,151,172,176
Conflict 29,35,88,104ff,137,167,
    172,189
Consensus 94ff,188
Construction 173,205
Convention 45
Cooperaton 31,35,57f,133,135,137,
    152,164
Creativity 71,74,83,157,159,196f,
    212,213,220
Culture 102,109,114,143,147
Cybernetics 138f,152

Decision 134,140,148,185
Determinism 59ff,67ff,129,166,199
Development 157
Differentiation, social 2,14,19f,
    21,96,97,148,151,157,171

Diffusion 83
Disequilibrium. See equilibrium
Dissipation 3f,209
Division of labour 134,142,162,209
DNA 5,33,58
Dysergy 155

Ecology 16,17,131,162
Embryology 53
Emergence 6,14,56f,66,129,135,
    141,196,202f,210f,220
Emotion. See affectivity
Energy 3,14,16,143
Entropy 3
Epigenesis 101-125
Epistemology 198ff,220ff
Epistemology, evolutionary 2,6,
    21,64
Equilibrium 60f,92,94,96,98,
    190,209f
Ethics. See morals
Evolution, general 172
Evolution, irreversibility of
    61,97
Evolution, open 55,65
Evolution, progressive 2,21,
    25,27,37,127,128f,142,151,157,
    170,173
Evolution, regressive 129
Evolution, socio-cultural 12,13,
    18ff,50,75,79ff,101-125,127,
    131,151,217f
Evolution, specific 172
Evolution, stage models of 111f,
    215
Evolution, theory of Iff,2,24ff,
    51ff,79ff,103ff,171ff
Evolutionary stabilized strategy (ESS)
    2,6,9,11,19
Expectation 6,88
Explanation 9ff,16,21,32,53,129
Fitness 14,20
Freedom 27,49ff,59ff,67
Function 15,54,88,131,143,155,209
Functionalism 3,96,157,216

Games theory 6,8,58

Genetics 3,29f
Genotyp 4
Goal 53ff,59,62,86,140,158,164

Habitualisation 71
Hierarchy 57,148,180f,193,199,
   201,204   ''
Historicism 62
History 50,104
Homeostasis 1
Hypercycle 56f,59

Ideology II,49,72,81f
Inborn release mechanism 16,17
Incest taboo 8,11,15ff,22,34
Individuation 213
Inequality. See stratification
Information 14,17,18,34,83,145,169,
   204,208
Instinct 1,13,28
Institution 11,12,15,17ff,88,89,
   92,98,115,134,150,158,172,176,
   177ff,193
Institutionalisation. See
   institution
Integration 148,151,166,186
Intention 8f,10,68
Interaction, social 2,163,173f,192f
Interest 34,36,38f,95
International relations 160ff
Interpenetration 214
Invention. See creativity

Justice, social 35f,91
Law 54,60
Learning 13,17,50,71,101,104,
   112,122,124,218f

Maladaption 145
Market 162,164,181ff
Meaning 6,10,11,20,90,219
Mechanics 4,68
Media of interaction 171-194
Mentalism 62ff
Metaphysics 12,55,79,203
Morality. See morals
Morals 23,26ff,33ff,37ff,42ff,
   72ff,102,111,144,185
Morphogenesis 4,11
Mutation 3,16,55,83,201

Niche 3,9,76,163,219

Norms. See rules

Ontogenesis 17
Opportunity 159f,168,177,182
Order 4,6,60,91,92,105
Organism 3f,5ff,68,130,133,195

Parasitism 163,195-224
Perception 5
Phenotyp 4,5ff
Phylogenesis 13
Physicalism 62ff,207
Planning 49,50,62,134
Plasticity, behavioural 16
Population 9,83,131,150
Predation 163
Probability 206,208
Punctualism 76
Purpose. See interaction

Rationality 9,18,65f,68,75,172
Rationalization 171
Reciprocity 163
Recursivness III
Reductionism 196f,201f,203,205
Regression 154,156,158
Relativism 40ff
Replication 188,192
Reproduction II,9,17,18,23,30,
   34,83,95,97,121,133,163,173f
Ressource 89,91,92,97,160
Rights. See rules
Ritual 105f
Rules 18,79-99,103ff,119ff,181
Rules, epigenetical 33ff

Science 189ff
Selection IIf,3f,7,12,14,18,23ff,
   29f,51ff,55,82ff,87,89,95f,101,
   128,130,131,132,142,163,173,201,
   221
Selective pressure 46,84,89,108
Self 214ff
Self-destruction 49
Self-maintainance. See self-
   organization
Self-organization III,4f,49,55ff,
   197
Self-planning. See self-
   organization
Self-production 101,112,122

Self-transformation III,91,96ff,
   103ff,112,120ff
Self-regulation. See self-
   organisation
Situation, social 171,174,183,210f
Social change. See self-
   transformation
Social darwinism 24ff,32,49f
Social sciences. See sociology
Social theory. See sociology
Socialization 11,17
Society 8,36,49ff,67ff,130,135
Society, primitive 8
Society, traditional 105ff
Sociobiology 2ff,5ff,28ff,31ff,
   41,44,76f
Sociology 1ff,32,37,51ff,85ff
Species 16,50,195
State formation 113ff,115ff,
   152,154
Stratification 14,27,36,92,99,
   109f,143
Structure 2,14,91
Structure, cognitive 66
Structure, social 12,211
Struggle for existence 29,163,
   195,201
Survival 9f,22,24,27,30,33,59,
   108,130,163,169,217
Symbiosis 163,195,215
Symbol 177,178,194,211f,213,219,
   221
Synergism 60,132ff,156,212
Synergism, negative 136
System, behavioural 2,9,17
System, political 142ff
Systems theory 51-77,130

Teleology 15,53,55,59,138
Teleonomy 55,131,142,167
Thermodynamics III,60,76
Time 3f,14
Trial and error 71,216,218

Value 10,18f,28,94,98,186
Variation 83f,90,101,163
Vitalism 3,52